AF628327

Birkhäuser

Frontiers in Mathematics

Advisory Editors

Laurent Saloff-Coste, Cornell University, Ithaca, NY, USA

Igor Shparlinski, The University of New South Wales, Sydney, NSW, Australia

Wolfgang Sprößig, TU Bergakademie Freiberg, Freiberg, Germany

This series is designed to be a repository for up-to-date research results which have been prepared for a wider audience. Graduates and postgraduates as well as scientists will benefit from the latest developments at the research frontiers in mathematics and at the "frontiers" between mathematics and other fields like computer science, physics, biology, economics, finance, etc. All volumes are online available at SpringerLink.

Baltic Institute of Mathematics Collection

Since 2012, the Baltic Institute of Mathematics has been organizing scientific events such as conferences, workshops, seminars, and schools. The main goal of these events is to facilitate networking between young and senior researchers to discuss the state-of-the-art and outstanding new results in mathematics, physics, and computer science, and their use in industry. The volumes in this collection are based on the Institute's events and are written in an accessible, pedagogical style for graduate students and early-career researchers.

Managing Editor
Maria Ulan, Baltic Institute of Mathematics

Editorial Advisory Board
Patrik Jansson, Chalmers University of Technology and University of Gothenburg
Sergiy Maksymenko, Institute of Mathematics of National Academy of Sciences of Ukraine
Peter J. Olver, University of Minnesota
Volodya Rubtsov, University of Angers

Noémie C. Combe · Philippe G. Combe ·
Hanna K. Nencka

Exploring Information Geometry

Recent Advances and Connections to Topological Field Theory

Birkhäuser

Noémie C. Combe
De Vinci Research Center
De Vinci Higher Education
Paris, France

Philippe G. Combe
Baltic Institute of Mathematics
Warsaw, Poland

Hanna K. Nencka
Baltic Institute of Mathematics
Warsaw, Poland

ISSN 1660-8046 ISSN 1660-8054 (electronic)
Frontiers in Mathematics
ISBN 978-3-032-23053-9 ISBN 978-3-032-23054-6 (eBook)
https://doi.org/10.1007/978-3-032-23054-6

Mathematics Subject Classification: 53B12, 62B10, 53D45

© The Editor(s) (if applicable) and The Author(s), under exclusive license to Springer Nature Switzerland AG 2026

This work is subject to copyright. All rights are solely and exclusively licensed by the Publisher, whether the whole or part of the material is concerned, specifically the rights of translation, reprinting, reuse of illustrations, recitation, broadcasting, reproduction on microfilms or in any other physical way, and transmission or information storage and retrieval, electronic adaptation, computer software, or by similar or dissimilar methodology now known or hereafter developed.
The use of general descriptive names, registered names, trademarks, service marks, etc. in this publication does not imply, even in the absence of a specific statement, that such names are exempt from the relevant protective laws and regulations and therefore free for general use.
The publisher, the authors and the editors are safe to assume that the advice and information in this book are believed to be true and accurate at the date of publication. Neither the publisher nor the authors or the editors give a warranty, expressed or implied, with respect to the material contained herein or for any errors or omissions that may have been made. The publisher remains neutral with regard to jurisdictional claims in published maps and institutional affiliations.

This book is published under the imprint Birkhäuser, www.birkhauser-science.com by the registered company Springer Nature Switzerland AG
The registered company address is: Gewerbestrasse 11, 6330 Cham, Switzerland

If disposing of this product, please recycle the paper.

Preface

This introductory text arises from a lecture given in Göteborg, Sweden, given by the first author and is intended for undergraduate students, as well as for any mathematically inclined reader wishing to explore a synthesis of ideas connecting geometry and statistics. At its core, this work seeks to illustrate the profound and yet natural interplay between differential geometry, probability theory, and the rich algebraic structures encoded in (pre-)Frobenius manifolds and related topics.

The exposition is structured into three principal parts. Part I provides a concise introduction to differential topology and geometry, emphasizing the role of smooth manifolds, connections, and curvature in the formulation of geometric structures. Part II is devoted to probability, measures, and statistics, where the notion of a probability space is refined into a geometric object, thus paving the way for a deeper mathematical understanding of statistical models. Finally, part III, we introduce (pre-)Frobenius manifolds and Gromov–Witten invariants, revealing their surprising connection to exponential families of probability distributions, learning methods and, discuss more broadly, their role in the geometry of information. At the end of those three parts the reader will find stimulating exercises.

By bringing together these seemingly distant disciplines, we aim to highlight the natural emergence of geometric structures in statistical theory. This work does not seek to be exhaustive but rather to provide the reader with a pathway into a domain of mathematics that is still in its formative stages, where many fundamental questions remain open. The first two parts of the text is accessible without requiring advanced prerequisites and should serve as an invitation to further exploration.

For pedagogical purposes, we have included comparison tables of various mathematical objects, footnotes with additional comments, and highlighted important remarks inside framed boxes. There are also illustrations designed to clarify ideas or concepts, as well as exercises with fully written solutions. The level of difficulty increases from one chapter to the next. The final part of the book is intended for readers who already have a solid foundation (Master's level) and for doctoral students.

Paris, France — Noémie C. Combe
Warsaw, Poland — Philippe G. Combe
Warsaw, Poland — Hanna K. Nencka

Acknowledgements

This research is part of the project No. 2022/47/P/ST1/01177 cofunded by the National Science Centre and the European Union's Horizon 2020 research and innovation programme under the Marie Skłodowska-Curie grant agreement no. 945339.

The first author extends heartfelt gratitude to the organizers of the Summer School and WorkshopWisła 23 and Maria Ulan for the invitation to lecture on information geometry, an opportunity that ultimately led to the creation of this book for students, researchers, and all those curious about this fascinating field. The first author is also grateful to Maxim Kontsevich for asking stimulating questions on the topic, that inspired deeper exploration and helped illuminate aspects of the subject that had remained unclear. This textbook is, in part, an effort to clarify these questions and provide a structured introduction to the topic. Finally, we would like to express all our gratitude to Yuri I. Manin, whose untimely death has saddened us all.

Contents

1 Introduction 1
1.1 Detailed Synopsis 2
1.2 Summary and Synthesis 4

Part I Manifolds, Topology and Geometry

2 Foundations in General Topology 9
2.1 Topological Space 10
2.1.1 Basis 11
2.1.2 Glossary of Topologies 12
2.2 Separated Space 13
2.3 Continuity 14
2.4 Topological Properties of Sets 15
2.4.1 Open and Closed Sets 15
2.4.2 Compactness 16
2.4.3 Connectedness 17
2.5 Homotopy 21
2.6 Disjoint Sum Topology 23
2.6.1 Disjoint Sum 23
2.6.2 Disjoint Sum Topology 23
2.7 More Exercises 25
2.8 Hints/Short Answers for Exercises 25

3 Topological and Modeled Manifolds 29
3.1 Definitions 30
3.2 Charts, Atlas 30
3.3 Modeled Manifolds 32
3.3.1 $\mathcal{E}$-Modeled Manifolds 32
3.3.2 Recollections on Locally Convex Spaces 34

3.3.3 Bridging Those Two Definitions ... 36
3.3.4 Relations to the Banach Space Implicit Function Theorem ... 37
3.4 Exercises ... 39

4 Differentiability and Gateaux Derivatives ... 41
4.1 Gateaux Directional Derivation: A Definition ... 42
4.1.1 Gateaux Derivative ... 42
4.1.2 Complex Case Specialization ... 43
4.1.3 Differentiable Manifolds ... 44
4.1.4 Differentiable Mappings ... 45
4.2 Manifolds Modeled on a Normed Vector Space ... 45
4.2.1 Real Manifolds, Coordinates, Charts ... 47
4.3 Tangent and Cotangent Spaces ... 48
4.3.1 Curves on a Manifold and Directional Derivatives ... 48
4.4 Tangent Vector—Tangent Space ... 49
4.4.1 Cotangent Vector, Cotangent Space ... 51
4.4.2 Implications of a Chart Change ... 53
4.4.3 Differential of a Function ... 54
4.5 Tensor at a Point of a Manifold ... 54
4.5.1 Universal Property ... 54
4.5.2 Multilinear Maps and Tensors of Type (r, s) ... 55
4.5.3 Tensor Products ... 56
4.5.4 Symmetry Properties of Tensors ... 58

5 Fiber Bundles ... 59
5.1 Definitions and Examples ... 59
5.1.1 Properties ... 61
5.1.2 Tangent Bundles and Frame Bundles ... 62
5.1.3 Tangent Bundle ... 63
5.1.4 Frame Bundle ... 63
5.2 Vector Fields, Tangent Bundles, Lie Algebras ... 64
5.2.1 Vector Field Lie Algebra ... 66
5.3 Cotangent Bundle—Covector Field ... 67
5.3.1 Cotangent Bundle ... 67
5.3.2 Covector Field ... 68
5.4 Action of a Diffeomorphism ... 69
5.4.1 Image of a Vector Field Under a Diffeomorphism ... 69
5.4.2 Image of a Cotangent Vector Field Under a Diffeomorphism ... 69
5.4.3 Tensor Field ... 70

6 Connections, Parallel Transport, and Sheafs 71
6.1 Linear Connection 72
6.1.1 Covariant Derivative of a Vector Field in the Direction of Y 77
6.1.2 The Covariant Derivative of a Tensor 78
6.2 Parallel Transport—Geodesics 80
6.2.1 Parallel Vector Along a Curve 80
6.2.2 Geodesics 81
6.3 Curvature 83
6.3.1 Vector Bundle—Version 83
6.3.2 Sheaf Version 84
6.3.3 Curvature Version 2 (Algebraic) 87
6.3.4 Sectional Curvature 93
6.4 Torsion in Differential Geometry 95
6.4.1 Torsion Tensor 95
6.4.2 Local Coordinates 96
6.4.3 Geodesic Deviations 97
6.4.4 Torsion Tensor 98
6.4.5 Relations to Curvature 98
6.4.6 Spencer Differential and Bianchi Identity 100

Part II Probabilities, Statistics, and Related Topics

7 Probability Theory—Measure Theory—Statistics 103
7.1 Key Information for Statistics 103
7.2 Probability 104
7.2.1 Probability Space 104
7.3 Examples 106
7.3.1 Flipping Coin Examples: How Biased Is It? 106
7.3.2 Other Examples 108
7.4 Measure 109
7.5 Measurable Function—Random Variables 110
7.5.1 Random Variables and Measurability 110
7.6 Dominating Measure 111
7.7 Information Geometry 113
7.7.1 Tangent Space to a Manifold of Probability Distributions 115
7.7.2 The Riemannian Metric 115
7.7.3 Parallel Transport in Statistical Manifolds 116
7.7.4 Covariant Derivative 116
7.7.5 Amari–Chentsov Tensor and α-Connections 116
7.7.6 Statistical Inference and α-Connections 120
7.8 Hints and Solutions to Exercises 123

8 Categorical Structures in Probability and Measure Theory 125
8.1 Philosophical Approach 126
8.1.1 What Is the Aim of Geometry? 126
8.1.2 Towards a Generalization of Klein's Plato-Figures 127
8.1.3 Categories as a Classifying Tool 128
8.2 Generalization of Klein's (Plato)-Figures/Congruences 131
8.2.1 Generalized Klein Figures 131
8.2.2 Markov Transformations 132
8.2.3 Generalized Klein Figures and Transformations 132
8.2.4 Klein's Congruences and the Universal Property 134
8.2.5 Concluding with Klein–Plato's Framework 138
8.3 Markov Transformations and Associated Subcategories 139
8.4 Dominated Families in the Category of Markov Morphisms 142
8.5 Dominated Families in the Category of Markov Morphisms 143
8.5.1 Conclusion and Remarks 144
8.5.2 Dominating Measures and Constructive Probabilities 144
8.6 Homogeneous Klein Geometries and Geodesic Structures 145
8.6.1 Geodesic Structures 145
8.6.2 Torsion, Curvature, and Flat Connections 146
8.6.3 Totally Geodesic Submanifolds 146
8.7 Equivariant Connections in the Category of Geometric Structures 147
8.7.1 Geodesic Midpoints and Natural Transformations 148

9 Categorical and Geometric Structures in Statistical Manifold Theory 151
9.1 Foundational Collections of Probability Measures 151
9.2 Categorical Framework for Statistical Manifolds 154
9.2.1 Types of Manifolds in Theory of Statistical Manifolds 156
9.2.2 Redundancy in Natural Coordinates 157
9.2.3 Invariant Local Coordinate Systems 158
9.2.4 Canonical Affine Coordinates in the Theory of Statistical Manifolds 160
9.2.5 Ceva Lines and Tangent Vector Fields 162
9.2.6 Affine Connections and Ceva Lines on Statistical Manifolds 164
9.3 Translation-Invariant Tensor Fields 166
9.4 Flat Connections 167
9.5 Exercises 167
9.5.1 Some Corrections 167

Part III Frobenius Manifolds and Information Geometry

10 Frobenius Manifolds ... 173
10.1 A Short Story of Frobenius Manifolds ... 173
10.1.1 Homological Mirror Symmetry (and More!) ... 174
10.2 Preliminary Notions: To Be (or Not) Affine ... 175
10.2.1 Crystallographic Groups and Fundamental Groups ... 176
10.2.2 2-Dimensional Affine Structures ... 177
10.2.3 Affine Structures on Complex Surfaces ... 178
10.2.4 n-Dimensional Affine Structures, $n \geq 3$ ... 178
10.2.5 Pre-Lie Structures ... 179
10.3 Flat Coordinates and Darboux–Egoroff Coordinates ... 180
10.4 Pre-Frobenius Manifolds ... 183
10.4.1 Essential Preliminary Ingredients ... 184
10.4.2 Multiplication Operation ... 184
10.4.3 Compatibility Relations ... 185
10.4.4 Potential Structure and Monge–Ampère Link ... 185
10.5 Frobenius Manifolds ... 187
10.5.1 Associative Pre-Frobenius Manifolds ... 187
10.5.2 Witten–Dijkgraaf–Verlinde–Verlinde Equation ... 188
10.5.3 Geometrization of Frobenius Structures ... 188
10.5.4 Structure Connections of Pre-Frobenius Manifolds ... 190
10.6 Emergence of Frobenius Structures on Statistical Manifolds ... 191
10.6.1 The Fundamental Tensors ... 191
10.6.2 The Hidden Multiplication ... 191
10.6.3 Summary of Key Structures ... 194
10.7 Semisimple Frobenius Manifolds ... 194
10.7.1 Definition and Characterizations ... 194
10.7.2 Canonical Decomposition in Semisimple Case ... 195
10.7.3 Splitting Structure ... 196
10.7.4 Geometric Interpretation: Deformations and Isomonodromy ... 198
10.7.5 Axiomatic Definition of Semisimple Frobenius Structures ... 200
10.7.6 Characterization of Semisimple Frobenius Structures ... 201
10.7.7 Structure Connection and Curvature ... 204
10.7.8 Semisimple Frobenius Manifolds and Web Structures ... 204
10.8 Solutions of Exercises ... 208

11 Unveiling the Hidden Geometry of Statistical Manifolds ... 211
11.1 Classical Projective Geometry ... 211
11.1.1 Projective Space ... 212
11.1.2 Normalization of Hypersurfaces and m-Pairs ... 217

11.1.3 Some Examples and Exercices ... 218
11.2 Generalization of Projective Spaces ... 219
11.2.1 Paracomplex Numbers ... 220
11.2.2 Module over a Paracomplex Algebra ... 221
11.2.3 Projective Paracomplex Space ... 222
11.3 Solutions to Exercise ... 223

12 How Frobenius structures, Paracomplex Spaces Meets Caph ... 225
12.1 Jordan Algebras and Mirrors ... 226
12.1.1 On Peirce's Reflections ... 226
12.1.2 Properties Related to Paracomplex Algebras ... 228
12.1.3 Geometry and Algebra of ∞-Dimensional Spaces in $\mathsf{Caph}(\Omega, \mathbf{S}, \mathbf{Z})$ ... 230
12.1.4 Paracomplex Geometry of $\mathsf{Caph}(\Omega, \mathbf{S}, \mathbf{Z})$ (Continuous Case) ... 232
12.1.5 Para-Kähler geometry of exponential families ... 237
12.2 Frobenius Structures on Submanifolds of $\mathcal{H}$ ... 239
12.2.1 Conclusion on Paracomplex and Frobenius Constructions ... 241

13 Novel Perspectives and Openings ... 245
13.1 Gromov–Witten Invariants: Classical and Statistical ... 245
13.2 Formal Frobenius Manifold ... 246
13.3 How to Link This to Statistics? ... 247
13.3.1 Gromov–Witten Invariants for Statistical Manifolds ... 248
13.4 Interpretation via the Log-Partition Function and GWS Invariants ... 250
13.4.1 Construction of GWS ... 251
13.5 Learning Geometrically ... 256
13.6 Final Word ... 261

References ... 263

Index ... 267

Introduction 1

A Note to the Reader

Welcome to this Introductory Textbook on Geometry and Information Geometry! Whether you are a student, researcher, or simply someone curious about this fascinating field, this book is designed to provide you with a clear and accessible entry point into the world of geometry and its modern extension into information geometry.

Geometry, in its many forms, has been a cornerstone of mathematics for centuries. With the advent of information theory and machine learning, new geometric structures have emerged that help us better understand complex probabilistic models, optimization, and data analysis. Information geometry is one such powerful framework, blending classical differential geometry with modern applications in probability and statistics.

This book is structured to guide you from the fundamentals of topology and differentiable manifolds to the more advanced concepts of probability geometry and Frobenius manifolds. The progression is designed to be intuitive, allowing you to build your knowledge step by step.

Throughout the chapters, you will find exercises to test your understanding, as well as solutions to some of the more challenging problems to aid your learning. To complement the material, you are encouraged to follow along with the accompanying YouTube video (see the QR code below), which aligns with this textbook and served as an inspiration for its creation.

We hope this book provides you with both insight and enjoyment as you embark on your journey into geometry and information geometry (Fig. 1.1).

© The Author(s), under exclusive license to Springer Nature Switzerland AG 2026
N. C. Combe et al., *Exploring Information Geometry*, Frontiers in Mathematics,
https://doi.org/10.1007/978-3-032-23054-6_1

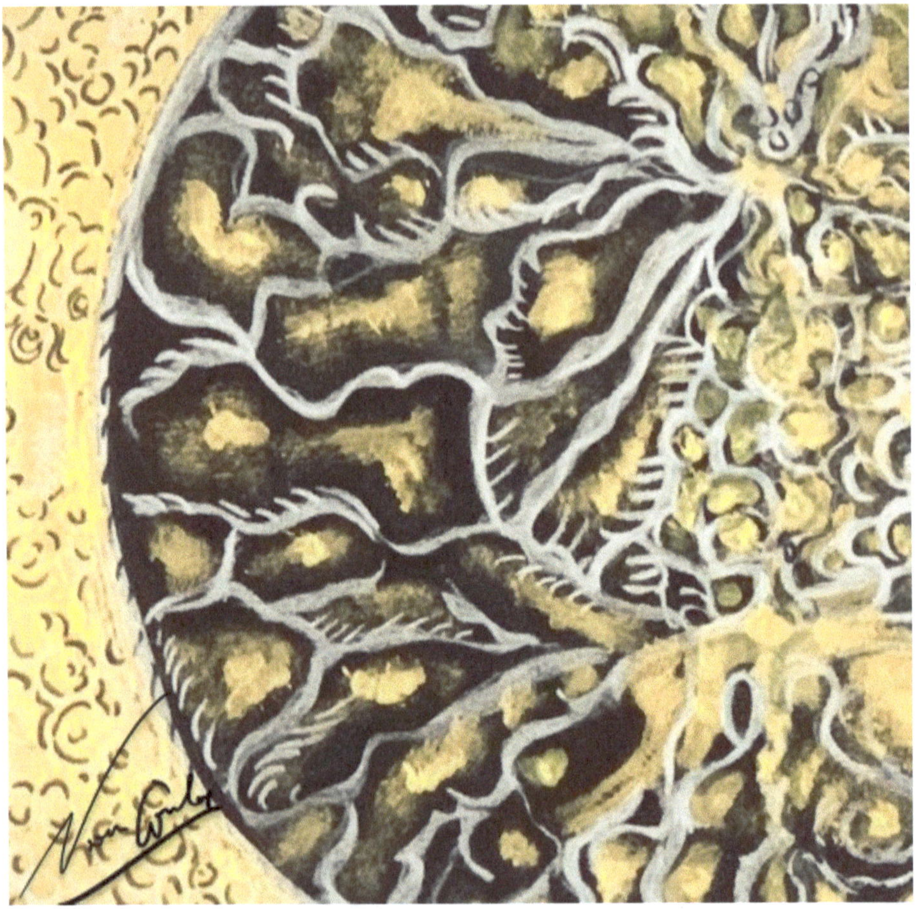

Fig. 1.1 A poem of geometric thoughts: an artistic vision, by Noémie Combe

Noémie C. Combe
Philippe G. Combe
Hanna K. Nencka

1.1 Detailed Synopsis

In what follows, we shall be concerned with a surprising fusion of ideas, namely, that of differential geometry and probability theory (or, more generally, statistics). At first glance, these two domains might appear unrelated, yet their synthesis gives rise to a highly rich and intricate mathematical structure, leading to the emergence of a relatively young field, known as the *geometry of information*. The relevance of this framework extends beyond pure mathematics, with applications in artificial intelligence (such as in language models like ChatGPT) and machine learning. Given the accelerating pace of developments in these areas, it is imperative to pursue the deeper mathematical underpinnings of this theory.

One of the great advantages of geometry is that it offers an intuitive and often visual means of understanding complex situations. This feature allows us to circumvent difficulties that might otherwise seem insurmountable. More precisely, at the heart of the geometry of information lies a fundamental object: the manifold of probability distributions, where the probability measures belong to a specific class of distributions.

To make matters precise, recall that a *probability space* consists of a triple $(\Omega, \mathcal{F}, P)$, where Ω is the sample space (the set of possible outcomes), $\mathcal{F}$ is a σ-algebra of measurable subsets of Ω, and P is a probability measure assigning to each event in $\mathcal{F}$ a real number in the interval $[0, 1]$. This object $(\Omega, \mathcal{F}, P)$ serves as a rigorous mathematical model for phenomena arising in nature. The challenge we face is to endow this structure with additional geometric data, thereby giving rise to a natural geometric space. However, this task is far from trivial.

A first step is to recognize that the set $\mathcal{F}$ carries a particular mathematical structure: that of a σ-algebra. The significance of σ-algebras lies in their closure properties: if $(\Omega, \mathcal{F})$ is a *measurable space*, then $\mathcal{F}$ is closed under countable unions, countable intersections, and complements. However, when seeking to introduce a geometric framework, it is often more convenient to work directly with probability distributions rather than the space $(\Omega, \mathcal{F})$ itself.

Probability distributions come in a vast array of different families. To clarify our discussion, let us recall the three principal classes of probability distributions:

(1) Discrete probability distributions, either with finite or infinite support.
(2) Absolutely continuous probability distributions, whose support may be:
    (a) A bounded interval.
    (b) An interval of length 2π (directional distributions).
    (c) A semi-infinite interval, $[0, \infty)$.
    (d) The entire real line.
(3) Probability distributions with variable support.

For our purposes, we restrict attention to the second class, focusing in particular on absolutely continuous distributions supported on intervals of length 2π, namely, *wrapped exponential distributions*.

A remarkable fact, which has emerged through careful analysis, is that probability distributions of this type possess a hidden geometric structure. The apparatus of differential geometry provides the natural language for uncovering and describing this structure. Objects such as connections, parallel transport, curvature, and flatness serve as fundamental tools in the construction of a geometric framework for probability distributions. This allows for the realization of a novel class of geometric spaces: *the manifolds of probability distributions*.

More concretely, consider a measurable space $(\Omega, \mathcal{F})$, where $\mathcal{F}$ is a σ-algebra on Ω. If we consider a family of parametrized probability distributions on $(\Omega, \mathcal{F})$, then the space of such distributions naturally inherits the structure of a manifold.

The study of these manifolds, particularly in the case of exponential families, leads to unexpected connections with *Topological Field Theory* (TFT). The relation between exponential families and TFT is not accidental; rather, it is deeply encoded in the mathematical structures underlying both domains. Indeed, Topological Field Theory is intimately linked to the celebrated *Witten–Dijkgraaf–Verlinde–Verlinde (WDVV) equations*, which in turn play a central role in the theory of *Frobenius manifolds*, developed and studied by Dubrovin, Manin, Kontsevich, and others.

A fundamental aspect of our approach, which renders manifest the deep relation between the manifold of probability distributions of exponential type and the (pre-)Frobenius manifold structures, lies in the delicate matter of choosing an appropriate system of coordinates. Indeed, the very possibility of perceiving this connection in an explicit and natural manner is contingent upon the identification of a privileged class of coordinates—ones that reflect, in their very definition, the intrinsic geometry of the underlying structures. The art, then, is not merely to introduce coordinates, but to do so in a manner that unveils (rather than obscures) the hidden algebraic and differential properties inherent in the space.

Thus, in certain cases, we are able to establish a deep and precise connection between the WDVV equations and the geometry of information. This, in turn, suggests that the study of probability distributions, when viewed through a geometric lens, is far richer than initially expected and might hold profound implications for both mathematical physics and information theory.

1.2 Summary and Synthesis

This introductory textbook is structured in three interrelated parts, each designed to build a solid foundation in modern mathematics and lead the reader toward the emerging field of information geometry.

Part 1: Manifolds, Topology, and Geometry

Chapter 2: Foundations of General Topology

The book begins with a modern treatment of general topology, reinterpreting Kuratowski's early ideas [Ku72] in a contemporary framework. Readers are introduced to essential topological concepts—such as open sets, continuity, convergence, and compactness—establishing the language and tools necessary for later discussions.

Chapter 3: Topological and Modeled Manifolds

Building on the groundwork of topology, the text moves into the realm of manifolds. Chapter 2 explains topological manifolds—spaces that locally resemble Euclidean space—while also presenting modeled manifolds based on S. Lang's influential works [La95, La99]. For additional reference, the text suggests consulting [Mi65] to cover aspects not fully developed in this book. This chapter serves as a bridge, transitioning from abstract topological spaces to concrete geometric structures.

Chapter 4: Differentiability and Gateaux Derivatives

Differentiability is explored next with a focus on Gateaux derivatives. This notion is particularly useful in infinite-dimensional settings, especially in probability theory where a norm may not be available. The text carefully contrasts Gateaux differentiability with the stronger concept of Fréchet differentiability, ensuring that students understand the subtleties and applications of both approaches.

Chapter 5: Fiber Bundles

Fiber bundles are introduced as indispensable tools in differential geometry. This chapter outlines the local-to-global perspective that fiber bundles offer, demonstrating how complex geometric structures can be understood by piecing together simpler, locally trivial components.

Chapter 6: Connections, Parallel Transport, and Sheafs

Delving deeper into the geometric framework, Chap. 5 covers connections, parallel transport, and covariant derivatives—key concepts for understanding how geometric data evolves along a manifold. Classical references such as [KobNo96, Sik71] provide further reading on these fundamental topics. Additionally, an introduction to sheafs is provided, offering a gentle entry point into their role in modern geometry, with [KaS90] recommended for a complete reference. This material lays the groundwork for applying geometric techniques within information geometry later in the book.

Part 2: Probabilities, Statistics, and Related Topics

Chapter 7: Basics of Probability and Statistics

The second part shifts focus to probability theory and statistics. It introduces standard definitions, theorems, and methodologies, employing familiar examples—such as coin flipping—to illustrate key concepts. The chapter also discusses the Radon–Nikodym derivative, linking measure theory with probabilistic reasoning. References such as [Bl95, Pa67] for measure theory and [Fe66] for statistics are suggested for further reading.

Chapters 8 and 9: Categorical and Geometrical Structures in Probability

These chapters are particularly innovative, blending philosophy with mathematics. Beginning with a discussion inspired by Klein's geometry and Plato's ideas, the text presents a novel perspective on how categorical structures naturally arise in the study of probability distributions. By considering manifolds of probability measures and the role of Markov kernels, the reader is guided through a modern generalization of classical geometry into a probabilistic and categorical setting. Key references for this section include [Am85, Am97, MoCh89, MoCh91-1, MoCh91-2].

Part 3: Frobenius Manifolds and Information Geometry

Chapters 10, 11, 12 , and 13: Frobenius Structures in Modern Research

The final part of the book covers Frobenius manifolds (geometric structures that elegantly encapsulate the interplay between algebra and geometry, coming from 2D Topological Field Theory) and Gromov–Witten invariants in the real of information theory. The text explains how Frobenius structures emerge naturally within the framework of information geometry and how Gromov–Witten invariants can provide interesting tools for the learning process. For foundational references on Frobenius manifolds and related topics, the book cites [D96, Ma99].

Chapter 13 integrates cutting-edge research from 2020 onward [CM20, CCN21], demonstrating the latest developments in the field. A notable highlight is the discussion of learning methods pioneered by researchers such as D. Ackley, G. Hinton, and T. Sejnowski [AHS85], showing how deep learning techniques relate to the broader mathematical framework of information geometry.

Conclusion This textbook provides a concise, accessible, and modern overview of several core areas of mathematics—topology, geometry, probability, and category theory—culminating in the study of information geometry. Through careful exposition and a judicious selection of topics, it equips students with the necessary tools to explore this young and rapidly evolving branch of mathematics, bridging classical theory with modern research and applications.

Part I

Manifolds, Topology and Geometry

Foundations in General Topology 2

General topology, also known as *point-set topology*, provides the foundational language for modern mathematics by studying the most fundamental properties of sets and their structures. Topology focuses on the intrinsic properties of spaces that remain unchanged under continuous transformations.

A topological space is a set X equipped with a topology—a collection of subsets called open sets that satisfy specific axioms and ensure consistency with notions of continuity and convergence. This framework generalizes many familiar mathematical spaces, including metric spaces and Euclidean spaces.

Let us mention some key concepts in general topology:

- Topological Spaces: A set equipped with a topology that defines continuity and convergence.
- Open and Closed Sets: Fundamental building blocks of topology that generalize intervals in real analysis.
- Basis and Sub-basis: Tools for constructing topologies using (minimal) generating sets.
- Continuity: A function between topological spaces is continuous if the preimage of every open set is open.
- Homeomorphisms: Structure-preserving maps that define when two spaces are topologically equivalent.
- Compactness: A property that allows to generalize the notion of a closed and bounded subset in an Euclidean space.
- Connectedness: A criterion telling whether a space can be separated into disjoint open sets.
- Separation Axioms: Conditions such as T_0, T_1, T_2 (Hausdorff), which control how points can be distinguished in a space.

© The Author(s), under exclusive license to Springer Nature Switzerland AG 2026
N. C. Combe et al., *Exploring Information Geometry*, Frontiers in Mathematics,
https://doi.org/10.1007/978-3-032-23054-6_2

These fundamental concepts serve as the basis for more advanced areas of mathematics, including analysis, geometry, algebraic topology, and functional analysis.

For readers already familiar with general topology, this section can be skipped. However, for those who need a refresher, the following discussion will provide a structured and intuitive approach to these essential topics before moving on to more advanced material.

2.1 Topological Space

Definition 2.1 (**Topological space**) Let X be a set and denote by $\mathcal{P}(X)$ the *power set*, that is, the set of all its subsets. A topology on X is a distinguished collection of subsets, $\mathscr{T} \subset \mathcal{P}(X)$, which we regard as specifying the admissible open sets. This collection satisfies the following axioms:

1. The set itself and the empty set belong to the topology:

$$\emptyset, X \in \mathscr{T}.$$

2. Any arbitrary union of open sets remains open:

$$\forall \{\mathcal{U}_i\}_{i \in I} \subset \mathscr{T}, \quad \bigcup_{i \in I} \mathcal{U}_i \in \mathscr{T}.$$

3. Any finite intersection of open sets remains open:

$$\forall \mathcal{U}_1, \ldots, \mathcal{U}_N \in \mathscr{T}, \quad \bigcap_{i=1}^{N} \mathcal{U}_i \in \mathscr{T}.$$

Thus, a topology provides the language of continuity: It defines what it means for a function to be continuous without reference to distances, relying only on the structure of open sets.

The elements of $\mathscr{T}$ are called the *open* sets of the topology; the conditions (1), (2), and (3) form the axioms of a topology.

To summarize, we can therefore also state this definition by saying that a topology is a collection of subsets of X, called open, which must verify that

1. the empty set and X are open;
2. any union of open sets is an open set;
3. a finite intersection of open sets is an open set.

We will sometimes write $(X, \mathscr{T})$ to specify that we are considering a set X equipped with its topology $\mathscr{T}$.

2.1.1 Basis

Definition 2.2 **(Basis)** A basis $\mathcal{B}$ for a topology $\mathscr{T}$ is a family of elements of $\mathscr{T}$ such that every $\mathcal{U} \in \mathscr{T}$ is the union of elements of $\mathcal{B}$.

We then refer to this as a topology generated by $\mathfrak{B}$.

An equivalent definition can be given. A basis $\mathfrak{B}$ for a topology $\mathscr{T}$ is a family of elements of $\mathscr{T}$ such that for each $x \in X$ and $\mathcal{U} \in \mathscr{T}$, with $x \in \mathcal{U}$, there exists $\mathcal{B} \in \mathfrak{B}$ such that $x \in \mathcal{B}$ and $\mathcal{B} \subset \mathcal{U}$.

Example 2.1 A basis for the usual topology of $\mathbb{R}$ is provided by the set of open intervals $\{]a, b[\mid a < b, a, b \in \mathcal{R}\}$.

An example of a topology widely used in practice is the metric topology.

Definition 2.3 **(Metric space)** A metric space is a set M endowed with a notion of distance, that is, a function $d : M \times M \to \mathbb{R}^+$ that satisfies:

1. $d(x, y) \geq 0, \ \forall x \in M.$
2. $d(x, y) = 0 \iff x = y, \ \forall x, y \in M.$
3. $d(x, y) = d(y, x), \ \forall x, y \in M.$
4. $d(x, z) \leq d(x, y) + d(y, z), \ \forall x, y, z \in M.$

The topology of M is generated by the open balls $B_r(a) = \{x \in M \mid d(a, x) < r\}$.

Exercise 2.1 The Cantor ternary set $\mathcal{C}$ is created by iteratively deleting the open middle third from a set of line segments. One starts by deleting the open middle third $(\frac{1}{3}, \frac{2}{3})$ from the interval $[0, 1]$, leaving two line segments: $[0, \frac{1}{3}] \cup [\frac{2}{3}, 1]$. One continues iterating a similar procedure for the remaining line segments. The Cantor set is constituted from all points in the interval $[0, 1]$ that are not deleted at any step in this infinite process. Is the Cantor set a metric space?

Definition 2.4 **(Neighborhood)**

1. A neighborhood of a point $x \in X$ is the set $\mathcal{N}(x)$ containing an open set that contains the point x.
2. A neighborhood of a set $A \subset X$ is the set $\mathcal{N}(A)$ containing an open set that contains A.

2.1.2 Glossary of Topologies

Definition 2.5 (Trivial and discrete topology)

▷ **Trivial Topology**: The trivial topology on a non-empty set X consists in taking as open sets the empty set $\emptyset$ and the entire space X.
▷ **Discrete Topology**: The discrete topology on a non-empty set X consists in taking as open sets the elements of $\mathcal{P}(X)$, the set of subsets of X.

Definition 2.6 (Coarser and finer topology)

▷ **Coarser Topology**: If $\mathscr{T}_1$ and $\mathscr{T}_2$ are two topologies on X such that $\mathscr{T}_1 \subset \mathscr{T}_2$, then $\mathscr{T}_1$ is said to be coarser than $\mathscr{T}_2$.
▷ **Finer Topology**: If $\mathscr{T}_1$ and $\mathscr{T}_2$ are two topologies on X such that $\mathscr{T}_1 \subset \mathscr{T}_2$, then $\mathscr{T}_2$ is said to be finer than $\mathscr{T}_1$.

- The coarsest topology is the trivial topology.
- The finest topology is the discrete topology.

Definition 2.7 (Induced topology on a subset) Let $(X, \mathscr{T}_X)$ be a topological space and $A \subset X$ be a subset. We define a topology $\mathscr{T}_A$ on A by setting:

$$\mathscr{T}_A = \{U \cap A \mid U \in \mathscr{T}_X\}.$$

In other words, we take as open sets of A the intersections of open sets of X with A.

Definition 2.8 (Quotient topology) Let $(X, \mathscr{T})$ be a topological space and $\mathcal{R}$ be an equivalence relation on X. Consider the canonical surjection

$$\begin{aligned} p : X &\to X/\mathcal{R}, \\ x &\mapsto [x] \end{aligned}$$

which maps an element $x \in X$ to its corresponding equivalence class $[x]$. The map p is called the canonical quotient map. The open sets of the quotient topology on $X/\mathcal{R}$ are formed from those subsets $\mathcal{V} \subset X/\mathcal{R}$ satisfying the condition that $\mathcal{V} = p^{-1}(\mathcal{U})$ is open. The set $\mathcal{V}$ is open in the quotient topology if and only if $\{x \in X : [x] \in \mathcal{V}\}$ is open in X.

2.2 Separated Space

In the study of topological spaces, separation properties play a crucial role in understanding the structure of a space and the behavior of functions defined on it.

Definition 2.9 (Hausdorff space—$\mathbf{T_2}$) A topological space $(X, \mathscr{T})$ is called Hausdorff (separated or $\mathbf{T_2}$) if for any pair of distinct points M and N, we can find two open sets $\mathcal{U}_M, \mathcal{U}_N$ with $M \in \mathcal{U}_M$, $N \in \mathcal{U}_N$ and $\mathcal{U}_M \cap \mathcal{U}_N = \emptyset$.

Definition 2.10 (**Normal space**) A topological space $(X, \mathscr{T})$ is normal if it is Hausdorff and if for any pair of disjoint closed sets F_1 and F_2, there exist two disjoint open sets $\mathcal{U}_1$ and $\mathcal{U}_2$ such that $F_1 \subset \mathcal{U}_1$ and $F_2 \subset \mathcal{U}_2$.

Theorem 2.1 *A Hausdorff space X is normal if and only if it satisfies the following condition:*

For every closed subset $F \subset X$ and every open subset $\mathcal{U}$ containing F, there exists an intermediate open subset $\mathcal{V}$ containing F satisfying

$$F \subset \mathcal{V} \subset \overline{\mathcal{V}} \subset \mathcal{U}.$$

Notice that here $\overline{\mathcal{V}}$ is properly defined in Sect. 2.4 under the terminology of *closure* of a set: $\overline{\{p\}} = \{p\}$.

Example 2.2 An open set $\mathcal{U} \subset \mathbb{R}^n$ is a Hausdorff space and moreover a normal space.

Definition 2.11 A topological space is $\mathbf{T_1}$ if for any two distinct points x and y, there exist open sets U and V such that:

- $x \in U, y \notin U$,
- $y \in V, x \notin V$.

In other words, each point is closed (its complement is open).

Exercise 2.2 Prove or disprove that any Hausdorff space is a $\mathbf{T_1}$ space. A $\mathbf{T_1}$ space is a space in which any set consisting of one point is closed.

⋆ Warning! There exist $\mathbf{T_1}$ space that are not $\mathbf{T_2}$.

Exercise 2.3 Give an example of a $\mathbf{T_1}$ space that is not $\mathbf{T_2}$.

2.3 Continuity

Definition 2.12 **(Continuity)** Consider two topological spaces $(X, \mathscr{T}), (X', \mathscr{T}')$. A map $f : (X, \mathscr{T}) \to (X', \mathscr{T}')$ is continuous if for any open set $\mathcal{U}' \in \mathscr{T}'$ its inverse image $f^{-1}(\mathcal{U}')$ is an open set of the topology $\mathscr{T}$.

Theorem 2.2 *1. A map $f : (X, \mathscr{T}) \to (X', \mathscr{T}')$ is continuous at the point $x_0 \in X$ if for every neighborhood $\mathcal{N}(f(x_0)) \subset X'$, there exists a neighborhood $\mathcal{N}(x_0) \subset X$ of $x \in X$ such that $f(x) \in \mathcal{N}(f(x_0))$, whenever $x \in \mathcal{N}(x_0)$.*
The map f is continuous on X if it is continuous at all $x \in X$.
2. A map f is continuous if and only if the sequence $\{f(x_\beta)\}$ converges to $f(x)$ when the sequence $\{x_\beta\}$ converges to x.

Exercise 2.4 Show that for any pair of continuous functions $f, g : X \to Y$, where X is a topological space and Y is a Hausdorff space, the set

$$\{x \in X \mid f(x) = g(x)\}$$

is closed in X.

Definition 2.13 **(Homeomorphism)** A homeomorphism between two topological spaces $(X, \mathscr{T}_X), (Y, \mathscr{T}_Y)$ is a continuous bijective map $h : (X, \mathscr{T}_X) \to (Y, \mathscr{T}_Y)$ whose inverse map is continuous.

Definition 2.14 **(Open map)** A map $f : X \to Y$ is called open if the image of any open set of X is an open set of Y.

Definition 2.15 **(Closed map)** A map $f : X \to Y$ between topological spaces is said to be closed if the subset

$$\Gamma_f = \{(x, f(x)) \mid x \text{ lies in a domain of } f\} \subset X \times Y,$$

The set Γ_f is closed in $X \times Y$. The set Γ_f is called the *graph* of f.

Note that open and closed maps are not necessarily continuous, as the notion of continuity is independent of openness and closedness.

2.4 Topological Properties of Sets

2.4.1 Open and Closed Sets

We will recall some definitions and theorems related to various properties of subsets of a topological space X.

- ▷ **Closed Subset**: A set $F \subset X$ is closed if it is the complement of an open set.
- ▷ **Limit Point**: A point $x \in X$ is a limit point of $A \subset X$ if every neighborhood $\mathcal{N}_x$ of x contains at least one point $a \subset A$ different from x.
- ▷ **Closure**: The closure $\bar{A}$ of $A \subset X$ is the union of A and all its limit points. It is the smallest closed set containing A.
- ▷ **Dense Subset**: The set $A \subset X$ is dense in X if $\bar{A} = X$.
- ▷ **Separable**: A topological space X is separable if it contains a countable dense subset.
- ▷ **Interior**: The interior $\mathring{A}$ of the set $A \subset X$ is the largest open set contained in A.
- ▷ **Nowhere Dense Set**: The set $A \subset X$ is nowhere dense in X if the interior of the closure of A is empty.

Definition 2.17 (**Baire Space**) A topological space is called a Baire space if the intersection of any countable family of dense open sets remains dense.

Equivalently, a topological space is Baire if the union of any countable collection of closed sets with empty interior also has an empty interior.

Complete metric spaces, as well as locally compact Hausdorff spaces, provide natural examples of Baire space.

- ▷ The set of integers is nowhere dense in the set of real numbers equipped with the usual topology.
- ▷ The set of real numbers whose decimal expansion only includes the digits 0 or 1 is nowhere dense in the set of real numbers.

Exercise 2.5 Is the set of rational numbers equipped with the subspace topology a Baire space? Write a proof.

Exercise 2.6 Is the set of real numbers equipped with standard topology a Baire space? Write a proof.

Theorem 2.3 *(***Baire's Category Theorem***)*

- ▷ *Every complete metric space is a Baire space.*
- ▷ *Every locally compact topological space is a Baire space.*

Theorem 2.4 *1. A subset $A \subset X$ is open if and only if it is a neighborhood of each of its points.*
2. A subset $A \subset X$ is closed if and only if it contains all its limit points.

2.4.2 Compactness

- ▷ **(Open) Covering**: A family $\mathcal{U}$ of open sets $U_\alpha \subset X$ is an open covering if $\bigcup_\alpha U_\alpha = X$.
- ▷ **(Open) Subcovering**: An open subcovering of a covering $\mathcal{U}$ is a subset of $\mathcal{U}$ which is itself an open covering.
- ▷ **Compact**: A subset $A \subset X$ is compact if it is compact as a subspace. In particular, every open covering of A has a finite subcovering.
  Note that we have not explicitly imposed the Hausdorff condition here; indeed, it can be omitted in many situations without affecting the validity of the statement.
- ▷ **Relatively Compact**: A subset $A \subset X$ is **relatively compact** if its closure $\bar{A}$ is compact.
- ▷ **Locally Compact**: A subset is locally compact if every point has a compact neighborhood. Notice that, Euclidean spaces are locally compact but not compact.
- ▷ **Paracompact**: A Hausdorff space is paracompact if every covering $\mathcal{U} = \{U_\alpha\}$ has a locally finite covering $\mathcal{V} = \{V_i\}$ such that, for every V_i there exists a U_β containing it ($V_i \subset U_\beta$).
- ▷ **Compactification**: A compactification of a topological space X is a pair (h, K) where K is a compact space and h is a homeomorphism of X onto a dense subspace of K.

Theorem 2.5 *1. A compact subspace of a Hausdorff space is necessarily closed.*
2. Every closed subspace of a compact space is compact.
3. **Bolzano–Weierstrass Theorem***: A Hausdorff space is compact if and only if every sequence has a convergent subsequence.*
4. **Heine–Borel Theorem** *The compact subsets of $\mathbb{R}^n$ are the closed bounded subsets of $\mathbb{R}^n$.*

The Heine–Borel theorem concerns finite-dimensional spaces, and it is not necessarily true for an arbitrary topological space. In particular, a closed bounded set with a non-empty interior of an infinite-dimensional normed vector space is never compact, for the norm topology.

Theorem 2.6 *Let K be a compact space. The image of K under a continuous map f is also compact.*

Corollary 2.1 *Given a compact space K, any continuous function on K attains on K a minimum and a maximum value.*

Exercise 2.7 Prove that the Cantor set is compact.

Exercise 2.8 Consider the set K of all functions $f : [0, 1] \to [0, 1]$ equipped with the Lipschitz condition, i.e., $|f(x) - f(y)| \leq |x - y|$ for all $x, y \in [0, 1]$. Consider on K the metric induced by the uniform distance

$$d(f, g) = \sup_{x\in[0,1]} |f(x) - g(x)|.$$

Prove that the space K is compact.

2.4.3 Connectedness

Definition 2.19 **(Connectedness)** A topological space X is said to be connected if it cannot be described as the union of two disjoint (non-empty) open sets. Otherwise X is said to be non-connected.

A subspace of a topological space is said to be connected if it is connected for the induced topology.

Exercise 2.9 Is the General Linear group $GL_n(\mathbb{R})$ (i.e., the space of square matrices of size $n \times n$ with non-null determinant) a connected space? Provide a proof.

Theorem 2.7 *For a topological space X the following conditions are equivalent:*

1. *X is connected.*
2. *X cannot be the union of two non-empty closed sets.*
3. *The subsets of X that are both open and closed are X and the empty set $\emptyset$.*
4. *X cannot be the union of two non-empty separable sets.*
5. *The continuous functions from $X \to \{0, 1\}$ are the constant functions.*

Definition 2.20 **(Locally Connected)** A topological space X is said to be locally connected if every neighborhood of every point $x \in X$ contains a connected neighborhood.

Example 2.4 1. The Euclidean line, the Euclidean plane, the cube I^n (without boundary) are connected and locally connected.

2. Any isolated point is a point of local connectivity.
3. The curve $y = \sin \frac{1}{x}$ on the interval $(0, 1]$ is connected but not locally connected. This is because the set contains the point $(0, 0)$, but it is not possible to connect the function to the origin.
4. Let $\mathbb{R}^2$ be equipped with the standard topology and let $K = \{\frac{1}{n} \mid n \in \mathbb{N}\}$. We call the comb space the set

$$C = (\{0\} \times [0, 1]) \cup (K \times [0, 1]) \cup ([0, 1] \times \{0\})$$

considered as a subspace of $\mathbb{R}^2$ equipped with the induced topology. The comb is a connected space that is not locally connected.

⋆ Warning! A space can be connected without being locally connected.

Exercise 2.10 Is the curve $y = \sin \frac{1}{x}$ on the interval $(0, 1]$ locally connected?

Exercise 2.11 Compute how many connected components has the real quartic surface:

$$f(u, v) = 3.2u - v^2 - \frac{1}{20} = 0$$

where $u = x^2 + y^2 + z^2$, $v = x^2y^2 + y^2z^2 + x^2z^2$.

Do the same for:

$$f(u, v) = 8u - v^2 - \frac{1}{20} = 0.$$

Definition 2.21 **(Covering space)** The covering of a topological space X is a pair $(\tilde{X}, f)$, where f is a continuous map from the topological space $\tilde{X}$ onto X, such that for every neighborhood $\mathcal{N}(x)$ of $x \in X$, the restriction of f to each connected component C_α of $f^{-1}(\mathcal{N}(x))$ is a homeomorphism from C_α to $\mathcal{N}(x)$. The connected components C_α are called sheets.

If $\mathcal{N}(x)$ is connected, then the sheets are uniquely determined up to homeomorphism.

Definition 2.22 **(Simply connected)** A topological space X is simply connected if X is connected and locally connected and any covering $(\tilde{X}, f)$ is isomorphic to the trivial covering (X, Id) where Id is the identity map.

▷ $\mathbb{R}$ is simply connected (roughly speaking it has no "holes" and every loop is shrunk to a point).
▷ The circle $S^1 = \mathbb{R}/\mathbb{Z}$ is not simply connected.

A universal covering space is the "simplest" simply connected space that covers a given space. Rigorously, it is defined as follows.

Definition 2.23 **(Universal covering)** $(\tilde{X}, f)$ is a universal covering of the space X if it is a covering and if $\tilde{X}$ is simply connected.

Example 2.6 A covering of S^1 is given by the pair $(\mathbb{R}, \pi)$, where π is the canonical projection given by

$$\pi(t) = \exp 2\pi \imath t$$

and t is a real parameter.

Exercise 2.12 Does the torus $S^1 \times S^1$ have a universal covering? If yes, give it explicitly.

Demonstrate that the Kähler torus $T^n = \mathbb{C}^n/\Lambda$ in $\mathbb{C}^n$ where Λ is a lattice in $\mathbb{C}^n$ (meaning that it is a discrete subgroup isomorphic to $\mathbb{Z}^{2n}$) has a universal covering space.

Definition 2.24 **(Locally simply connected)** A topological space X is locally simply connected if every point $x \in X$ has at least one simply connected neighborhood.

Example 2.7 The sphere S^2 is locally simply connected, since every point on the sphere has a simply connected neighborhood. For every point on S^2, one can choose a small open neighborhood around that point, homeomorphic to an open disk in $\mathbb{R}^2$.

Exercise 2.13 We consider a space which consists of infinitely many circles of decreasing radius, all tangent to a single point. This is called a Hawaiian earring. Prove that this space is not locally simply connected.

Definition 2.25 **(Isomorphism of covering)** Two coverings $(\tilde{X}_1, f_1)$ and $(\tilde{X}_2, f_2)$ are isomorphic if

1. $\varphi : \tilde{X}_1 \to \tilde{X}_2$ is a homeomorphism.
2. $f_1 = f_2 \circ \varphi$.

Definition 2.26 **(Fundamental group)** Let X be a space that admits a universal covering $(\tilde{X}, f)$. The group of homeomorphisms φ of $\tilde{X}$ onto itself such that $f \circ \varphi = f$ is called the fundamental group of $\tilde{X}$.

Since two universal coverings are isomorphic, so are the fundamental groups. The corresponding abstract group is called the fundamental group of X.

Exercise 2.14 Compute the fundamental groups of the following topological spaces:

- Sphere S^n.
- Torus T^2.
- Real projective space $\mathbb{R}P^n$.
- Hawaiian earring.

Theorem 2.8 *1. A connected and locally connected space has a universal covering.*
2. Two universal coverings are isomorphic.
3. Any covering of a universal covering is trivial.

Definition 2.27 (Path-connected)

1. A topological space X is path-connected, if for any two points a and b in X, there exists a continuous path connecting those points; that is, there exists a continuous function $\gamma : [0, 1] \to X$ such that $\gamma(0) = a$ and $\gamma(1) = b$.
2. X is locally path-connected (or pathwise connected) if for each point $x \in X$ and each neighborhood $\mathcal{N}(x)$ there exists a neighborhood $\mathcal{U}(x) \subset \mathcal{N}(x)$ that is path-connected.

Theorem 2.9 *If a topological space is path-connected (locally path-connected), then it is connected (locally connected).*

⋆ Warning! The converse statement is false.

Example 2.8 Consider the topologist's sine curve, which is constituted from the graph of the function $y = \sin(\frac{1}{x})$, where $x \in (0, 1]$ and the vertical line segment $\{0\} \times [-1, 1]$. The space is connected however it is not path-connected. Indeed, there is no continuous path connecting a point on the vertical segment to a point on the sine curve. In particular, any neighborhood around any point on $\{0\} \times [-1, 1]$ contains infinitely many disconnected components of the sine curve.

Exercise 2.15 Provide an other example where the space is connected but not path-connected.

2.5 Homotopy

The notion of homotopy is a purely topological notion which allows to consider classes of topological objects up to some (homotopical) relation.

To give an everyday life example, up to homotopy, a mug is equivalent to a doughnut (because they are both homotopy equivalent to a torus).

Definition 2.28 **(Free homotopic paths)** Two paths $\gamma_1, \gamma_2 : [0, 1] \to X$ are free homotopic if there exists a continuous map $F : [0, 1] \times [0, 1] \to X$ such that $F(t, 0) = \gamma_1(t)$ and $F(t, 1) = \gamma_2(t)$.

Definition 2.29 **(Path homotopic paths)** Two paths $\gamma_1, \gamma_2 : [0, 1] \to X$ are path homotopic if there exists a continuous map $F : [0, 1] \times [0, 1] \to X$ such that $F(t, 0) = \gamma_1(t)$, $F(t, 1) = \gamma_2(t)$ and $F(0, s) = a$, $F(1, s) = b$ for all $t, s \in [0, 1]$.

Example 2.9 In the figure below, we draw a pair of homotopic planar curves in red and blue, respectively. These curves are the level curves of the real and imaginary parts of a pair of complex polynomials.

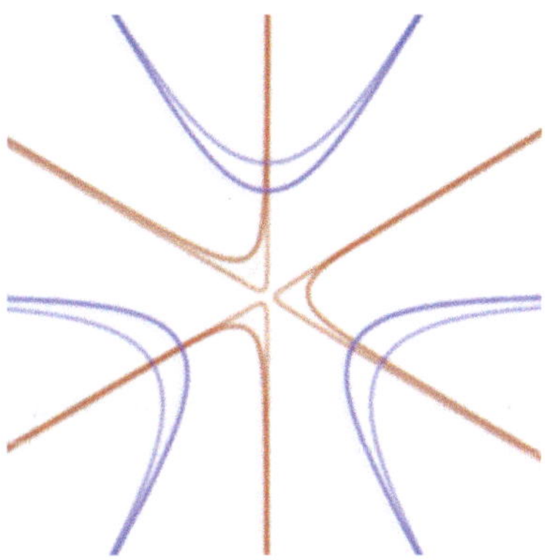

Can you tell the degree of those curves? Give (approximately) the equations of those polynomials.

Definition 2.30 **(Homotopic map)**

1. Two continuous maps $f, g : X \to Y$ are homotopic if there exists a continuous map $F : X \times [0, 1] \to Y$ such that $F(x, 0) = f(x)$ and $F(x, 1) = g(x)$ for all $x \in X$.
2. If A is a subset of X, two continuous maps $f, g : X \to Y$ are homotopic with respect to A if there exists a continuous map $F : X \times [0, 1] \to Y$ such that $F(x, 0) = f(x)$ and $F(x, 1) = g(x)$ for all $x \in X$ and $F(a, t) = f(a)$ for all $a \in A$ and $t \in [0, 1]$.

Theorem 2.10 *If X is path-connected and locally path-connected, it is simply connected if every path γ in X is homotopic to the constant map.*

The notion of homotopy between two functions allows to define an equivalence relation between topological spaces.

Definition 2.31 **(Homotopically equivalent spaces)** Two spaces X and Y are said to be homotopically equivalent (or "of the same homotopy type") if there exist two continuous maps $f : X \to Y$ and $g : Y \to X$ such that:

- $g \circ f$ is homotopic to Id_X, the identity on X;
- $f \circ g$ is homotopic to Id_Y, the identity on Y.

Definition 2.32 **(Contractible space)** A space is called contractible if it is homotopically equivalent to a point.

This definition is equivalent to saying that its identity map is homotopic to a constant map.

Example 2.10 The space $\mathbb{R}^n$ is contractible.

Theorem 2.11 *Two homeomorphic topological spaces are homotopically equivalent*

> Warning! The converse is false, as shown by the following examples:

- ▷ Except from the point itself, contractible spaces are not homeomorphic to the point.
- ▷ A circle is homotopically equivalent to $\mathbb{C}^\star$, that is, a plane without a point, but it is not homeomorphic to it. A circle without two points is not connected, while a plane without three points is.
- ▷ The real interval, the disk in the plane or the ball in $\mathbb{R}^3$, of radius 1 (open or closed) are all contractible, therefore homotopically equivalent. However, pairwise they are non-homeomorphic (by analogous arguments).

2.6 Disjoint Sum Topology

2.6.1 Disjoint Sum

Definition 2.33 (Disjoint sum) The disjoint sum of an indexed family of sets $\{E_i\}_{i\in I}$ is defined by:

$$\coprod_{i\in I} E_i = \bigcup_{i\in I} \{(x,i) \mid i \in I,\ x \in E_i\}. \tag{2.1}$$

Note: This definition of the disjoint sum allows to take into account the case where the $E_i \cap E_j \neq \emptyset$. Indeed, the elements of the disjoint sum are ordered pairs (x, i). Here $i \in I$ serves as an auxiliary index that indicates from which E_i the element x came from. Each of the E_i sets is canonically isomorphic to the set $\tilde{E}_i = \{(x,i) : x \in E_i\}$. So, if we define $\varphi_j : E_j \to \coprod_{i\in I} E_i$ by $\varphi_j(x) = (x, j)$, where $j \in I$ then φ_j is a bijection from E_j onto $\tilde{E}_j = \{(x,i) \mid x \in E_j\}$ and the $\tilde{E}_j$ are disjoint, even if the E_i are not.

$$\coprod_{i\in I} E_i = \bigcup_{i\in I} \{(x,i) \mid i \in I,\ x \in E_i\} = \bigsqcup_{i\in I} \{(x,i) \mid i \in I,\ x \in E_i\}.$$

Usually to state that the union is a disjoint union we replace $\bigcup$ by $\bigsqcup$.

2.6.2 Disjoint Sum Topology

Definition 2.34 (Disjoint sum topology) The disjoint sum topology on $\coprod_{i\in I} E_i$ is defined by

$$\mathcal{T} = \left\{ \mathcal{U} \subset \coprod_{i\in I} E_i \mid \varphi_i^{-1}(\mathcal{U}) \in \mathcal{T}_i\ \forall\, i \in I \right\}.$$

Proposition 21 *(Fundamental properties of the disjoint sum topology)*

1. *The set* $\left\{ \mathcal{U} \subset \coprod_{i\in I} E_i \mid \varphi_i^{-1}(\mathcal{U}) \in \mathcal{T}_i\ \forall\, i \in I \right\}$ *defines a topology.*
2. *The maps* $\varphi_j : X_j \to \coprod_{i\in I} E_i$ *are continuous, open, and closed; they are in fact homeomorphisms on*
$$\varphi_j(X_j) = \left\{ (x, j) \in \coprod_{i\in I} E_i \mid x \in E_j \right\}.$$
3. *Let* Y *be a topological space and* $f : \coprod_{i\in I} E_i \to Y$ *be a map. Then*
$$f \text{ is continuous } \Leftrightarrow f \circ \varphi_j : E_j \to Y \text{ is continuous}, \forall\, j \in I.$$
4. *The disjoint sum topology is the finest among those for which the* φ_j*, for all* $j \in I$ *are continuous.*

Proof 1. It is clear that $\varnothing$ and $\coprod_{i\in I} E_i$ belong to the topology.
Axioms (2) and (3) follow because these are topologies and because:

(a) For a family of subsets of $\coprod_{i\in I} E_i$:

$$\varphi_j^{-1}\left(\bigcup_{\alpha} \mathcal{U}_\alpha\right) = \bigcup_{\alpha} \varphi_j^{-1}(\mathcal{U}_\alpha) \quad \text{and} \quad \varphi_j^{-1}\left(\bigcap_{\alpha\in A} \mathcal{U}_\alpha\right) = \bigcap_{\alpha\in A} \varphi_j^{-1}(\mathcal{U}_\alpha)$$

so both unions and finite intersections are open in E_j if each $\mathcal{U}_\alpha$ is open.

(b) Let $\mathcal{U} \subseteq \coprod_{i\in I} E_i$ such that $\varphi_j^{-1}(\mathcal{U})$ is open in E_j. Then $\mathcal{U}$ is open by definition of the coproduct topology.

2. **Closedness**: Let $F \subseteq E_j$ be closed. Define

$$A = \coprod_{i\in I} E_i \setminus \varphi_j(F).$$

Then:

$$\varphi_i^{-1}(A) = \begin{cases} E_j \setminus F & \text{if } i = j, \\ E_i & \text{otherwise.} \end{cases}$$

Since $E_j \setminus F$ is open in E_j and E_i is open in itself for $i \neq j$, each $\varphi_i^{-1}(A)$ is open. Hence A is open in $\coprod_{i\in I} E_i$, so $\varphi_j(F)$ is closed.

3. **Homeomorphism onto Image**: Each φ_j is continuous, open, injective, and closed onto its image, hence a homeomorphism onto $\varphi_j(E_j)$.

4. **Universal Property**:

- ($\Rightarrow$) If $f : \coprod_{i\in I} E_i \to Y$ is continuous, then for each j, the composition $f \circ \varphi_j : E_j \to Y$ is continuous as a composition of continuous maps.
- ($\Leftarrow$) Suppose each $f \circ \varphi_j$ is continuous. Let $V \subseteq Y$ be open. Then:

$$\varphi_j^{-1}(f^{-1}(V)) = (f \circ \varphi_j)^{-1}(V)$$

is open in E_j for all j. Hence, by definition of the coproduct topology, $f^{-1}(V)$ is open in $\coprod_{i\in I} E_i$, so f is continuous.

5. Let $\mathcal{T}'$ be a topology on $\coprod_{i\in I} E_i$ such that each φ_j is continuous. For any open $U \in \mathcal{T}'$, the set $\varphi_j^{-1}(U)$ is open in E_j. Thus, U is open in the coproduct topology and $\mathcal{T}' \subseteq \mathcal{T}$. Hence, the coproduct topology is the finest topology making each φ_j continuous.

2.7 More Exercises

1. Prove or disprove that every Hausdorff space $\mathbf{T_1}$ that is a space in which every set consisting of one point is closed.
2. Give an example of $\mathbf{T_1}$ space which is not Hausdorff space.
3. For each pair f, g of continuous maps of a topological space X in a Hausdorff space Y the set $\{X \in X : f(x) = g(x)\}$ is closed in space X.

2.8 Hints/Short Answers for Exercises

Exercise 2.1 and 2.7 The Cantor set is a metric space. The Cantor set is an uncountable set with Lebesgue measure zero. As the complement of a union of open sets, it is a closed subset of the real numbers and, consequently, a complete metric space. Furthermore, since it is totally bounded, the Heine–Borel theorem ensures that it is compact.

Exercise 2.2 Let p be a given point of the space X. By hypothesis X is a $\mathbf{T_2}$ space. Therefore, it follows that every point $x \neq p$ belongs to an open set G_x, that does not contain p and thus that

$$X \setminus \{p\} = \cup_{x \neq p} G_x.$$

This implies that $X \setminus \{p\}$ is open, in other words $\{p\}$ is closed.

Exercise 2.3 Consider the following topological space:

$$X = \left\{0\right\} \cup \left\{\frac{1}{n} \;\middle|\; n \in \mathbb{N}^*\right\}.$$

We equip X with a topology defined as follows:

- The open sets that do not contain the point $\{1\}$ are precisely those inherited from the standard topology on $\mathbb{R}$. For example $\{\frac{1}{2}, \frac{1}{3}\}$ is open.
- Any open set that contains the point $\{1\}$ must necessarily be of the form $U = X \setminus F$, where F is a finite subset of X not containing 1. For example, $U = X \setminus \{\frac{1}{2}, \frac{1}{3}\}$.

Let us check that X is a $\mathbf{T_1}$ Space. To show that X is $\mathbf{T_1}$, we need to prove that every singleton set $\{x\}$ is closed. Consider:

- For $x = 0$: the space $X \setminus \{0\}$ is open.
- For $x = \frac{1}{k}$: the space $X \setminus \{\frac{1}{k}\}$ is open because it can be written as $X \setminus F$, where $F = \{\frac{1}{k}\}$ is finite.

Therefore, every singleton is closed, and X is $\mathbf{T_1}$.

Let us check if X is $\mathbf{T_2}$.

- Any open set U containing 0 must be of the form $X \setminus F$, where F does not contain 0. Since $X \setminus F$ is infinite, U must contain infinitely many points of the form $\frac{1}{n}$.
- Any open set V containing 1 must also be of the form $X \setminus F$, where F is finite and does not contain 1. This set V will also contain infinitely many points of the form $\frac{1}{n}$. Thus U and V cannot be disjoint because they both contain infinitely many points $\frac{1}{n}$. Therefore, 0 and 1 cannot be separated by disjoint open sets and X is not $\mathbf{T_2}$.

To summarize,

- the space X is $\mathbf{T_1}$, because every singleton is closed;
- the space X is not $\mathbf{T_2}$, because the points 0 and 1 cannot be separated by disjoint open sets.

Exercise 2.4 is enough to show that the set

$$A = \{x \in X : f(x) \neq g(x)\}$$

is open. For every $x \in A$ there exists in Y two open sets U and V such that $f(x) \in U$ and $g(x) \in V$ and their intersection is empty. The set given by $f^{-1}(U) \cap g^{-1}(V)$ is the neighborhood of the point x and contained in A. Therefore, A is an open set.

Exercise 2.5 Let us consider the example of a non-Baire space: the set of rational numbers with the subspace Topology.

The set of rational numbers is countable. In particular, we can write

$$\mathbb{Q} = \cup_{q \in \mathbb{Q}} \{q\}.$$

Each singleton $\{q\}$ is closed in $\mathbb{Q}$ (subspace topology) but has empty interior: Thus, $\{q\}$ is nowhere dense. Therefore, $\mathbb{Q}$ is a countable union of nowhere dense sets and $\mathbb{Q}$ cannot be a Baire space.

Exercise 2.6 The real numbers $\mathbb{R}$, equipped with the standard Euclidean topology, forms a Baire space.

A space is Baire if the intersection of countably many dense open subsets is dense. By the Baire Category Theorem, every complete metric space is a Baire space. Therefore, the conclusion follows.

Exercise 2.8 This is an application of the Arzela–Ascoli theorem.

Exercise 2.9 No. This is due to the fact that $GL_n(\mathbb{R})$ splits into two disjoint subsets: the set of matrices with strictly positive determinant and the set of matrices with strictly negative determinant.

Exercise 2.10 The curve $y = \sin \frac{1}{x}$ on the interval $(0, 1]$ is connected but not locally connected nor arc-connected. As $x \to 0^+$, $\lim_{x \to 0^+} \frac{1}{x}$ grows without bound, causing $\sin \frac{1}{x}$ to oscillate infinitely between the values -1 and 1. By definition, a space is locally connected if every neighborhood of any point contains a connected open neighborhood. In the case of

$y = \sin \frac{1}{x}$, no matter how small a neighborhood U of $(0, y)$ is chosen, the graph within U consists of infinitely many disconnected components (due to the oscillations). Thus, there is no connected open neighborhood around $(0, y)$.

Exercise 2.11 The first quartic has eight non-compact connected components, while the second has only one. This can be shown by leveraging the fact that these quartics are invariant under the Coxeter group CB_3, which corresponds to the symmetries of the cube. By decomposing space into Coxeter chambers, the quartic can be analyzed within a single fundamental domain. Applying an appropriate change of variables, $x_i^2 \mapsto X_i$ where $i \in \{1, 2, 3\}$ within this domain reduces the problem to studying degree-2 surfaces, defined in a cone contained in a positive octant (for instance the cone delimited by the equations $\{x_1 = 0\}, \{x_2 = 0\}, \{x_3 = 0\}$, and $\{x_i = x_j\}$ where $i \neq j$. Since the classification of quadrics is well understood, this approach simplifies the analysis. See [C18] for an entire classification of such quartics.

Exercise 2.12 The universal covering space of the torus is the Euclidean plane $\mathbb{R}^2$, with the covering map: $\pi : \mathbb{R}^2 \to T^2, \pi(x, y) = (\exp 2\pi \imath x, \exp 2\pi \imath y)$.

The Kähler torus (that is a torus equipped with a Kähler structure) admits a universal covering. In fact, the universal covering of a Kähler torus is the same as the universal covering of a standard torus, because the Kähler structure does not affect the underlying topology. The map π projects each point $z \in \mathbb{C}^n$ to its equivalence class $z + \Lambda$ in the quotient space $\mathbb{C}^n/\Lambda$. This is a smooth, holomorphic map that respects the complex and Kähler structures. The space $\mathbb{C}^n$ is simply connected because it is contractible.

Exercise 2.13 The Hawaiian earring is a classic example of a space that is not locally simply connected. It consists of infinitely many circles of decreasing radius, all tangent to a single point. While each circle is itself simply connected, the entire space fails to be locally simply connected at the point of tangency because any neighborhood of that point contains infinitely many circles, making it impossible to find a simply connected neighborhood.

Exercise 2.14 Fundamental groups.

- $\pi_1(S^1) \cong \mathbb{Z}$.
- $\pi_1(S^n) \cong 0$, if $n \geq 2$.
- $\pi_1(S^1 \times S^1) \cong \mathbb{Z} \times \mathbb{Z}$.
- $\pi_1(\mathbb{R}P^1) \cong \mathbb{Z}$.
- $\pi_1(\mathbb{R}P^n) \cong \mathbb{Z}/2\mathbb{Z}$.
- Hawaiian earring: uncountable.

Exercise 2.15 There are many examples. We can take the example of the deleted comb space. The comb space is a subspace of $\mathbb{R}^2$ which looks like a comb. A comb space is defined by the set:

$$\{0\} \times [0, 1]) \cup (K \times [0, 1]) \cup ([0, 1] \times \{0\}$$

where $K = \{\frac{1}{n} \mid n \in \mathbb{N}^*\}$. However, the aim of this exercise is to create your own example, using your imagination.

Topological and Modeled Manifolds

3

A topological manifold is a fundamental object in mathematics, serving as an abstraction of spaces that locally resemble Euclidean space but exhibit additional properties. Manifolds provide the natural setting for geometry and topology, forming the foundation for different disciplines in mathematics.

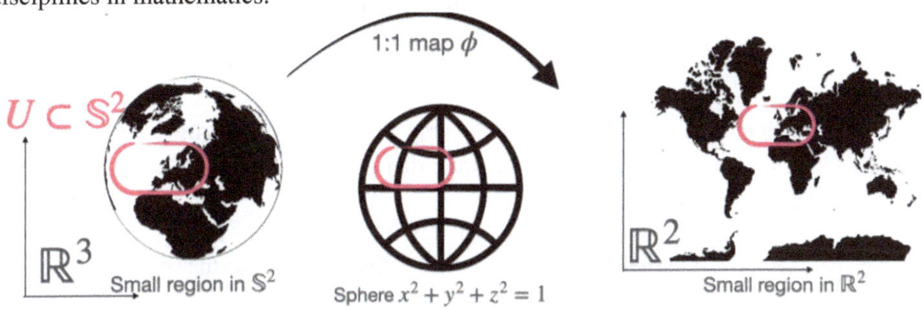

Formally, an n-dimensional topological manifold is a Hausdorff topological space that is locally homeomorphic to $\mathbb{R}^n$. This means that around every point, there exists a neighborhood that behaves like an open subset of the Euclidean space, allowing us to use an Euclidean space type of intuition, while still permitting the existence of global structures (namely curvature, holes, or any other topological feature).

Topological manifolds can be studied with some extra structures:

- Differentiable manifolds, where smooth functions and derivatives can be defined, leading to differential geometry.
- Riemannian manifolds, which are equipped with a Riemannian metric.
- Lie Groups: These are topological spaces that also possess a group structure, playing a crucial role in symmetry and mathematical physics.

© The Author(s), under exclusive license to Springer Nature Switzerland AG 2026

N. C. Combe et al., *Exploring Information Geometry*, Frontiers in Mathematics,
https://doi.org/10.1007/978-3-032-23054-6_3

The simplest examples of topological manifolds include familiar spaces such as the circle S^1, the sphere S^n the torus T^n, as well as more sophisticated objects like projective spaces.

3.1 Definitions

Definition 3.1 (Topological manifolds) Let $\mathcal{M}$ be a set, and let $(\mathcal{E}_i)_{i\in I}$ be a family of Hausdorff topological vector spaces. Consider a collection $\mathcal{A}$ of pairs $\{(\mathcal{U}_i, \varphi_i)\}_{i\in I}$, indexed by some set I, where each $\mathcal{U}_i$ is a subset of $\mathcal{M}$ and each φ_i is a mapping associated to it, subject to the following axioms:

($\mathcal{M}$**1**) The sets $\mathcal{U}_i$ form an open cover of $\mathcal{M}$:

$$\mathcal{M} = \bigcup_{i\in I} \mathcal{U}_i, \quad \mathcal{U}_i \subset \mathcal{M}.$$

($\mathcal{M}$**2**) Each φ_i is a bijection between $\mathcal{U}_i$ and an open subset of $\mathcal{E}_i$:

$$\varphi_i : \mathcal{U}_i \to \varphi_i(\mathcal{U}_i) \subset \mathcal{E}_i.$$

($\mathcal{M}$**3**) For each pair of indices $i, j \in I$, the transition maps

$$\varphi_j \circ \varphi_i^{-1} : \varphi_i(\mathcal{U}_i \cap \mathcal{U}_j) \to \varphi_j(\mathcal{U}_i \cap \mathcal{U}_j)$$

define homeomorphisms between the respective open subsets of $\mathcal{E}_i$ and $\mathcal{E}_j$.

($\mathcal{M}$**4**) Unless otherwise specified, we assume that each $\mathcal{E}_i$ is a locally convex Hausdorff topological vector space.

The data $(\mathcal{M}, \mathcal{A})$ thus defines a *topological manifold* modeled on the spaces $\mathcal{E}_i$ (Fig. 3.1).

3.2 Charts, Atlas

Definition 3.2 (Chart) Let $\mathcal{M}$ be a topological manifold. Each pair $(\mathcal{U}_i, \varphi_i) \in \mathcal{A}$, where $\mathcal{U}_i$ is an open subset of $\mathcal{M}$ and φ_i is a mapping of $\mathcal{U}_i$ onto an open subset of a topological vector space, is called a *chart* (or *coordinate system*) of $\mathcal{M}$.

If a point $\mathrm{M} \in \mathcal{M}$ lies in $\mathcal{U}_i$, we say that $(\mathcal{U}_i, \varphi_i)$ is a chart at M.

Definition 3.3 (Atlas) An *atlas* on a manifold $\mathcal{M}$ is a collection of charts $\{(\mathcal{U}_i, \varphi_i)\}_{i\in I}$ such that:

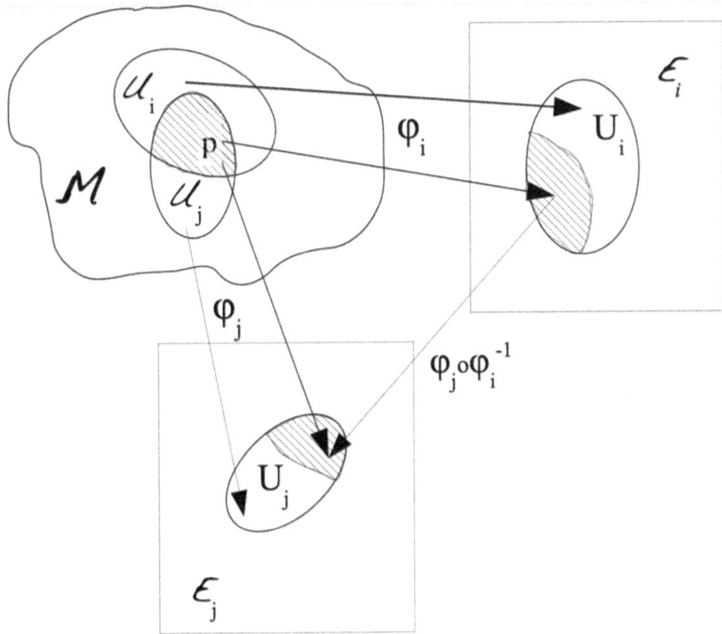

Fig. 3.1 Charts and atlas

(A1) The sets $\mathcal{U}_i$ cover $\mathcal{M}$:

$$\mathcal{M} = \bigcup_{i \in I} \mathcal{U}_i.$$

(A2) The transition maps

$$\varphi_j \circ \varphi_i^{-1} : \varphi_i(\mathcal{U}_i \cap \mathcal{U}_j) \rightarrow \varphi_j(\mathcal{U}_i \cap \mathcal{U}_j)$$

are homeomorphisms for all i, j for which $\mathcal{U}_i \cap \mathcal{U}_j \neq \emptyset$.

It is natural to assume that the topology on $\mathcal{M}$ is given a priori. In many cases, one requires that each chart φ_i is a homeomorphism onto its image. As a first consequence of this definition, we obtain the following structural result:

Proposition 1 *One can give to $\mathcal{M}$ a topology in a unique way such that each $\mathcal{U}_i$ is open, and the φ_i are topological isomorphisms.*

Remark 3.1 The condition that $\mathcal{M}$ is Hausdorff is not necessary. This condition plays no role in the formal development of manifold.

However, in practical applications, $\mathcal{M}$ is Hausdorff.

Exercise 3.1 Prove that the Special Linear group $SL_n(R)$ is a smooth submanifold of $\mathbb{R}^{n\times n}$.

Exercise 3.2 Is the set $\{x, y \in \mathbb{R}^2 \mid xy = 0\}$ a submanifold of $\mathbb{R}^2$?

3.3 Modeled Manifolds

Hausdorff Condition and Separation Axioms

In what follows, we impose the Hausdorff condition on all manifolds under consideration. Furthermore, any construction performed subsequently—such as products, tangent bundles, or fibered structures—will be required to yield spaces that remain Hausdorff. This ensures a well-behaved topological framework in which geometric operations preserve separation properties.

3.3.1 $\mathcal{E}$-Modeled Manifolds

In the formulation given above, particularly in condition $\mathcal{M}2$, no global assumption was imposed on the structure of the topological vector spaces $\mathcal{E}_i$ used as local models. In particular, we did not require that all $\mathcal{E}_i$ be the same, nor that there exist continuous linear isomorphisms between them.

However, one of the most natural and frequently encountered cases in practice is that in which there exists a fixed topological vector space $\mathcal{E}$ such that each $\mathcal{E}_i$ is isomorphic to $\mathcal{E}$, allowing a uniform model for the local structure of $\mathcal{M}$. In such a setting, one can regard $\mathcal{M}$ as being locally modeled on a single space $\mathcal{E}$, which provides a more rigid but also more manageable framework for various constructions.

This remark leads us to the notion of modeled manifold (Fig. 3.2).

Definition 3.4 (Modeled manifold) Let $\mathcal{M}$ be a topological space. An *$\mathcal{E}$-atlas* on $\mathcal{M}$ consists of a collection of charts $\{(U_i, \varphi_i)\}_{i\in I}$, where:

1. $\{U_i\}_{i\in I}$ is an open cover of $\mathcal{M}$.
2. Each $\varphi_i : U_i \to \mathcal{E}_i$ is a homeomorphism onto its image, where $\mathcal{E}_i$ is a topological space.
3. All the spaces $\mathcal{E}_i$ are homeomorphic to a fixed topological space $\mathcal{E}$, i.e., there exist homeomorphisms $h_i : \mathcal{E}_i \to \mathcal{E}$.

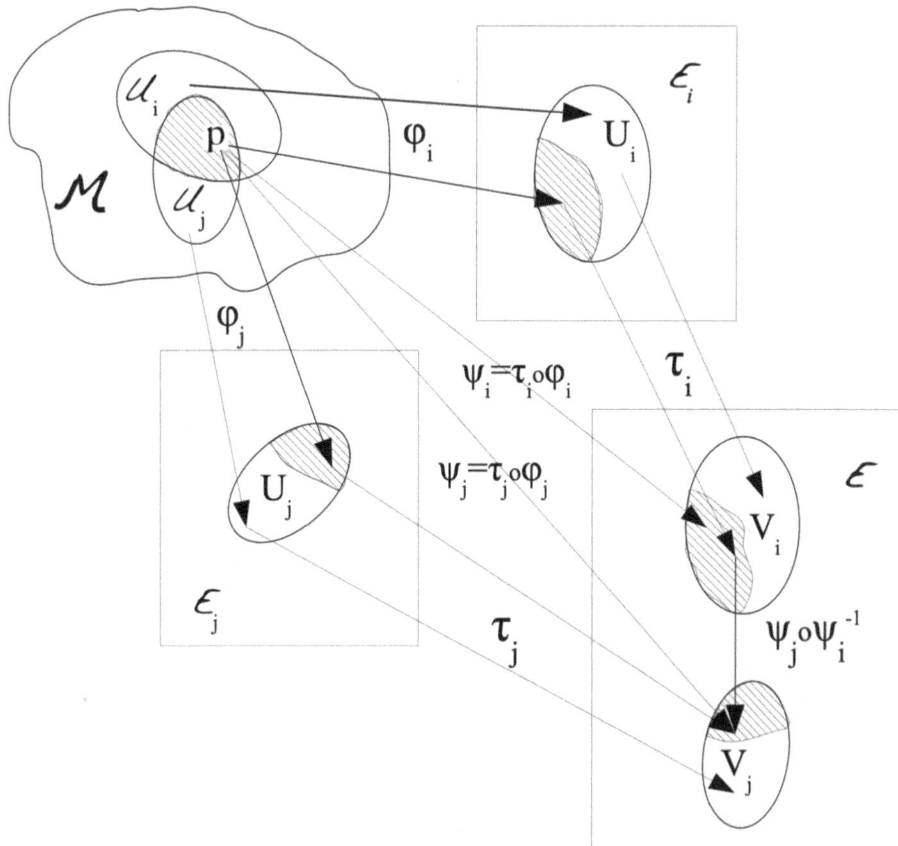

Fig. 3.2 Modeled manifold

If an $\mathcal{E}$-atlas is given, we say that $\mathcal{M}$ is an *$\mathcal{E}$-modeled manifold* (or simply an *$\mathcal{E}$-manifold*).

If the transition maps (i.e., maps of the type $\varphi_j \circ \varphi_i^{-1}$) satisfy additional compatibility conditions (such as differentiability, smoothness, or analyticity), then the $\mathcal{E}$-manifold is endowed with some extra structure.

In the following, we primarily focus on modeled manifolds whose model space is a vector space that admits a well-defined differentiable structure. This is the case for manifolds modeled on Banach spaces or Hilbert spaces, where a differentiable structure can be naturally defined. However, we are also interested in a generalization of normed spaces, namely locally convex topological vector spaces.

Example 3.1 A Hilbert manifold is a manifold modeled on a Hilbert space, which is a complete inner product space. Hilbert manifolds generalize finite-dimensional smooth manifolds to infinite dimensions.

- **Model Space**: Let H be a separable Hilbert space (e.g., $L^2(\mathbb{R})$).
- **Topological Space**: $\mathcal{M}$ is a Hausdorff and second-countable topological space.
- **Charts**: $\mathcal{M}$ is equipped with an atlas $\{(U_\alpha, \phi_\alpha)\}$ where:
  - $U_\alpha \subseteq \mathcal{M}$ is open.
  - $\phi_\alpha : U_\alpha \to \phi_\alpha(U_\alpha) \subseteq H$ is a homeomorphism onto an open subset of H.
- **Transition Maps**: For any two charts (U_α, ϕ_α) and (U_β, ϕ_β) with $U_\alpha \cap U_\beta \neq \emptyset$, the map

$$\phi_\beta \circ \phi_\alpha^{-1} : \phi_\alpha(U_\alpha \cap U_\beta) \to \phi_\beta(U_\alpha \cap U_\beta)$$

  is smooth (infinitely Fréchet differentiable) between open subsets of H.
- **Local Structure**: At each $p \in \mathcal{M}$, the tangent space $T_p\mathcal{M}$ is isomorphic to H.

Exercise 3.3 Consider a smooth finite-dimensional manifold $\mathcal{M}$. The loop space $L\mathcal{M}$ is the space of all smooth maps:

$$\gamma : S^1 \to \mathcal{M},$$

where S^1 is the unit circle. Is the loop space $L\mathcal{M}$ a modeled manifold?

Exercise 3.4 Let $\mathcal{M}$ and $\mathcal{N}$ be finite-dimensional smooth manifolds. Let $C^k(\mathcal{M}, \mathcal{N})$ be the space of C^k-differentiable maps from $\mathcal{M}$ to $\mathcal{N}$. Prove that this space can be given the structure of a Banach manifold modeled on a Banach space.

3.3.2 Recollections on Locally Convex Spaces

Let us recall that a **locally convex space** X is a vector space over $\mathbb{K}$, a (sub)field of the complex numbers (it can be $\mathbb{C}$ itself or $\mathbb{R}$ for instance).

A locally convex space is defined either in terms of convex sets or equivalently in terms of seminorms. In fact, a topological vector space X is said to be locally convex if it verifies one of the following two equivalent properties. We start with the convex set definition.

Definition 3.5 (Convex sets definition) A topological vector space X is said to be locally convex if there exists a neighborhood basis (that is, a local base) at the origin, consisting of balanced convex sets.

We elaborate on those two notions of convexity and of balanced sets.

1. A subset $C \in X$ is called *convex* if for all $x, y \in C$, and $0 \leq t \leq 1$ we have
$$tx + (1 - t)y \in C.$$
2. A subset $C \in X$ is called *balanced* if for all $x \in C$ and scalars $s \in \mathbb{K}$, the condition $|s| \leq 1$ implies that $sx \in C$.
3. A subset which is both convex and balanced is said to be *absolutely convex*.
4. A convex subset $C \in X$ is called a *cone* (we assume here that the underlying field is ordered) if for all $x \in C$ and $t \geq 0$ we have $tx \in C$.
5. A convex subset $C \in X$ is called *absorbent* if for every $x \in X$ there exists $r > 0$ such that
$$x \in t \cdot C$$
for all $t \in \mathbb{K}$, and $|t| > r$. The set C can be scaled out by any "large" value to absorb every point in the space.

In any topological vector space, every neighborhood of the origin is absorbent.

A second possible viewpoint on this notion can be achieved using seminorms. A *seminorm* on X is a function
$$\mathbf{p} : X \to \mathbb{R}$$
such that:

1. **Non-negativity**: $\mathbf{p}$ is nonnegative or positive semidefinite, i.e.,
$$\mathbf{p}(x) \geq 0, \quad \text{for all} \quad x \in X.$$
2. **Scaling Property**: $\mathbf{p}$ is positive homogeneous or positive scalable:
$$\mathbf{p}(sx) = |s| \cdot \mathbf{p}(x), \quad \text{for every scalar} \quad s \text{ and } x \in X.$$
So, in particular, $\mathbf{p}(0) = 0$.
3. **Subadditivity**: $\mathbf{p}$ is subadditive and it satisfies the triangle inequality:
$$\mathbf{p}(x + y) \leq \mathbf{p}(x) + \mathbf{p}(y), \quad \text{for all} \quad x, y \in X.$$

Definition 3.6 (Seminorms definition) A topological vector space X over a field $\mathbb{C}$ or $\mathbb{R}$ is said to be locally convex if there exists a family $\mathcal{P}$ of *seminorms* on X. Let $\{\mathbf{p}_i\}_{i \in I}$ be a family of seminorms on X, where I is an index set. Each seminorm $\mathbf{p}_i : X \to \mathbb{R}$ satisfies the properties of a seminorm (non-negativity, absolute homogeneity, and subadditivity).

Remark 3.2 Although the definition in terms of a neighborhood base gives a better geometric picture and intuition, the definition in terms of seminorms is easier to work with, in practice.

Exercise 3.5 Consider the space of continuous functions $C(\mathbb{R})$ on $\mathbb{R}$. Verify whether the family of functions $\{\mathbf{p}_n\}_{n\in\mathbb{N}}$, given by:

$$\mathbf{p}_n(f) = \sup_{x\in[-n,n]} |f(x)|$$

forms a family of seminorms. Is $C(\mathbb{R})$ locally convex? Do the $\{\mathbf{p}_n(f)\}$ generate a locally convex topology?

3.3.3 Bridging Those Two Definitions

As mentioned earlier, both definitions are equivalent. We outline a sketch of the proof showing this equivalence.

> The equivalence of those two definitions follows from a construction known as the *Minkowski functional* or Minkowski gauge. The key feature of seminorms which ensures the convexity of their ε-balls is the *triangle inequality*.

For an absorbing set C such that: if $x \in C$ then $t \cdot x \in C$, whenever $0 \leq t \leq 1$, let us define the Minkowski functional of C to be

$$\mu_C(x) = \inf\{r > 0 : x \in r\,C\}.$$

From this definition, it follows that μ_C is a seminorm if C is balanced and convex (it is also absorbent by assumption).

Conversely, given a family of seminorms, the sets

$$\left\{x : \mathbf{p}_{\alpha_1}(x) < \varepsilon_1, \ldots, \mathbf{p}_{\alpha_n}(x) < \varepsilon_n\right\}$$

form a base of convex absorbent balanced sets.

Let us mention some interesting properties.

- If $\mathbf{p}$ is positive definite (which states that if $\mathbf{p}(x) = 0$ then $x = 0$), it implies that $\mathbf{p}$ is a norm.

While in general seminorms *need not* be norms, there is an analogue of this criterion for families of seminorms and separatedness, defined below.

- If X is a vector space and $\mathcal{P}$ is a family of seminorms on X then a subset $\mathcal{Q} \subset \mathcal{P}$ is called a base of seminorms for $\mathcal{P}$ if for all $\mathbf{p} \in \mathcal{P}$ there exists a $\mathbf{q} \in \mathcal{Q}$ and a real number $r > 0$ such that $\mathbf{p} \leq r\mathbf{q}$.
- If X is a real vector space and C a convex subspace, if $x \in C$, then C contains the line segment between x and $-x$.
- If X is a complex vector space and C a convex subspace, for any $x \in C$ convex space, contains the disk with x on its boundary, centered at the origin, in the one-dimensional complex subspace generated by x.

3.3.4 Relations to the Banach Space Implicit Function Theorem

A *locally convex topological vector space* is a topological vector space in which every neighborhood of 0 contains an open neighborhood U of 0 such that, for all $x, y \in U$ and $0 \leq t \leq 1$,

$$t \cdot x + (1 - t) \cdot y \in U.$$

This property turns out to be essential in the context of the implicit function theorem, for Banach spaces.

The latter result, known as the **Banach Space Implicit Function Theorem**, can be stated as follows:

Let X, Y, and Z be Banach spaces and let $\mathcal{U}$ be an open subset of $X \times Y$. Suppose that $f : \mathcal{U} \to Z$ is a continuously differentiable (C^1) mapping such that $f(a, b) = 0$ for some $(a, b) \in \mathcal{U}$ and the partial derivative $D_y f(a, b)$ is a linear isomorphism from Y onto Z.

Then, there exists:

- an open set $\mathcal{W} \subset X$ with $a \in \mathcal{W}$;
- an open set $\mathcal{V} \subset \mathcal{U} \subset X \times Y$ with $(a, b) \in \mathcal{V}$;
- a continuously differentiable (C^1) mapping $g : \mathcal{W} \to Y$,

such that

$$(x, y) \in \mathcal{V},\ f(x, y) = 0 \iff x \in \mathcal{W},\ y = g(x).$$

In other words, the implicit equation $f(x, y) = 0$ has for $x \in W$ a solution $y = g(x)$ of class C^1 such that $(x, y) \in \mathcal{V}$.

This solution is unique in an open set $\mathcal{W}' \subset \mathcal{W}$.

WARNING: It is crucial to note that *not* every locally convex topological vector space admits a differentiable structure in a meaningful way. Nevertheless, a particular class of locally convex topological vector spaces, known as *Fréchet spaces*, plays a fundamental role in various areas of mathematics, such as:

- Functional analysis.
- Partial differential equations.
- Mathematical statistics.
- Infinite-dimensional geometry.

Fréchet spaces are characterized by their complete metric topology induced by a countable family of seminorms, which enables differential calculus.

A *Fréchet space* X is a locally convex topological vector space that is metrizable and complete, meaning:

- Its topology is induced by a translation-invariant metric d (or equivalently, by a countable family of seminorms $\{\|\cdot\|_k\}_{k\in\mathbb{N}}$)
- Every Cauchy sequence (x_n) with respect to d converges to some $x \in X$.

For normed vector spaces X, Y, the *Fréchet derivative* of a function $f : \mathcal{U} \to Y$ (where $\mathcal{U} \subseteq X$ is open) at $x \in \mathcal{U}$ is the unique bounded linear operator $A : X \to Y$ satisfying:

$$\lim_{h\to 0} \frac{\|f(x+h) - f(x) - Ah\|_Y}{\|h\|_X} = 0.$$

When it exists, A is denoted $Df(x)$. In finite-dimensional spaces ($X = \mathbb{R}^m$, $Y = \mathbb{R}^n$), this coincides with the classical derivative and is represented by the Jacobian matrix $J_f(x)$.

Remark 3.3 1. The definition requires X, Y to be *normed* spaces. For Fréchet spaces (which may not be normable), differentiation requires seminorms:

$$\lim_{h\to 0} \frac{\|f(x+h) - f(x) - Ah\|_k}{\|h\|_m} = 0 \quad \forall k, m.$$

2. The operator A must be *bounded* (continuous) and linear—a stronger requirement than Gateaux differentiation.
3. In $\mathbb{R}^n$, $Df(x)h = J_f(x)h$ (matrix–vector product).

3.4 Exercises

Exercise 3.3 The loop space $L\mathcal{M}$ is modeled on the Hilbert space $H = L^2(S^1, \mathbb{R}^n)$, where $n = \dim(\mathcal{M})$. Each loop γ can be locally approximated by functions in $L^2(S^1, \mathbb{R}^n)$. For a fixed loop $\gamma \in L\mathcal{M}$, a chart around γ can be constructed using the exponential map on $\mathcal{M}$. For a small neighborhood U of γ, the chart maps U to an open subset of H.

Exercise 3.4 This space can be given the structure of a Banach manifold, modeled on a Banach space of sections of a vector bundle. The model space is the Banach space $C^k(\mathcal{M}, \mathbb{R}^n)$ where $n = \dim(\mathcal{N})$. For a fixed map $f \in C^k(\mathcal{M}, \mathcal{N})$, a chart around f can be constructed using the exponential map on $\mathcal{N}$. Specifically, for a small neighborhood U of f, the chart maps U to an open subset of $C^k(\mathcal{M}, \mathbb{R}^n)$.

Exercise 3.5 Yes, each $\boldsymbol{p}_n$ forms a seminorm. The family $\{\boldsymbol{p}_n\}_{n\in\mathbb{N}}$ generates a locally convex topology on $C(\mathbb{R})$. In this topology, a sequence of functions f_k converges to f if and only if $\boldsymbol{p}_n(f_k - f) \to 0$, for every $n \in \mathbb{N}$.

Exercise 3.1 This exercise is slightly more advanced since this statement can be shown using the Regular Value Theorem, which we have not considered in this book. We will use this exercise as a possibility to mention and illustrate this theorem!

1. - A first step is to identify the space of all $n \times n$ matrices to the Euclidean space $\mathbb{R}^{n^2}$.
   - The determinant function det is a smooth (infinitely differentiable) function since it is a polynomial in the entries of the matrix.
   - By definition, the special linear group is a set of matrices of size $n \times n$ where the determinant is equal to 1. In other words, this is the preimage of the value 1, under the determinant function.
2. The Regular Value Theorem states that if

$$f : \mathcal{M} \to \mathcal{N}$$

   is a smooth map between two smooth manifolds and $y \in \mathcal{N}$ is a regular value of f, then $f^{-1}(y)$ is a smooth submanifolds of $\mathcal{M}$. In our case, we have $\mathcal{M} = \mathrm{Mat}_n(\mathbb{R}) = \mathbb{R}^{n^2}$, $\mathcal{N} = \mathbb{R}$, and $f = \det$. One needs to prove that 1 is a regular value of det. This can be done by showing that for every matrix $A \in \mathrm{Sl}_n(\mathbb{R})$ the derivative $d(\det)_A$ is surjective.
3. We now compute the derivative of the determinant. The derivative of the determinant function at a matrix $A \in \mathrm{Mat}_n(\mathbb{R})$ is given by the following formula:

$$d(\det)_A : \mathrm{Mat}_n(\mathbb{R}) \to \mathbb{R}, \quad d(\det)_A(H) = \mathrm{tr}(\mathrm{adj}(A) \cdot H),$$

   where adj stands for the classical adjoint of the square matrix A (it is defined as the transpose of the cofactor matrix of A) and tr is the trace operator. If $A \in \mathrm{Sl}_n(\mathbb{R})$, then the formula becomes

$$d(\det)_A(H) = \mathrm{tr}(A^{-1} \cdot H),$$

since $\text{adj}(A) = A^{-1}$ if $A \in \text{Sl}_n(\mathbb{R})$.

4. Finally, the last step is to show the surjectivity of the derivative. In other we need to demonstrate that for any real number $c \in \mathbb{R}$, there exists a matrix $H \in \text{Mat}_n(\mathbb{R})$ such that:
$$\text{tr}(A^{-1}H) = c.$$
An easy example of such matrices is given for $H = cA$. Indeed,
$$\text{tr}(A^{-1}H) = \text{tr}(A^{-1}(cA)) = c \cdot \text{tr}(I_n) = c \cdot n.$$
Therefore, $d(\det)_A$ is surjective for all $A \in \text{Sl}_n(\mathbb{R})$.
5. Applying the theorem leads to the conclusion.

Exercise 3.2 The equation $f(x, y) = xy = 0$ in the Euclidean plane $\mathbb{R}^2$ describes a union of the two coordinate axes: the x-axis (corresponding to $y = 0$) and y-axis corresponding to ($x = 0$). This set is not a submanifold of $\mathbb{R}^2$ in the usual sense of differential geometry.

1. Observe that there exists a *singular point* at $(0, 0)$. The set
$$S = \{x, y \in \mathbb{R}^2 \mid xy = 0\}$$
consists of two connected components (the x-axis and the y-axis), which intersect precisely at the origin. Intuitively, a smooth submanifold of $\mathbb{R}^2$ must locally resemble a smooth curve (or a single point). Away from the origin, the set S consists of a pair of smooth lines, but the origin, the submanifold condition fails.
2. To verify this, we compute the partial derivatives of the function $f(x, y) = xy$:
$$\frac{\partial f(x, y)}{\partial x} = y \quad \text{and} \quad \frac{\partial f(x, y)}{\partial y} = x.$$
At $(0, 0)$, both derivatives vanish:
$$\frac{\partial f(x, y)}{\partial x}(0, 0) = 0 \quad \frac{\partial f(x, y)}{\partial y}(0, 0) = 0.$$
This confirms that $(0, 0)$ is indeed a singular point.

This illustrates an example of a singular algebraic variety, where the smooth parts (the axes away from the origin) are manifolds, but the overall structure is not a manifold due to the singularity at the origin.

Differentiability and Gateaux Derivatives 4

In this chapter, we undertake a comprehensive investigation of the notion of differentiability, within an extended framework. This expands beyond elementary calculus to encompass both differentiable functions and the rich structure of differentiable manifolds. Building upon this foundation, we systematically develop the essential constructions of tangent vectors and tangent spaces, alongside their dual counterparts, cotangent spaces. These concepts not only underpin the analytical machinery of differential geometry but also enable far-reaching applications across mathematics—from topology and dynamical systems—and physics, particularly in the geometric formulation of classical mechanics, general relativity, and gauge field theories.

More precisely, the notion of differentiability serves in mathematics in:

1. Local Linear Approximations

    - Differentiability allows functions, curves, or maps to be approximated locally by linear objects (e.g., tangent lines, Jacobian matrices). This is used for instance in optimization, where one considers extrema via critical points. This is also used in machine learning.

2. Smoothness (and Other Important Properties) of Manifolds

    - Differentiability defines the notion of "smoothness" of a manifold but also permits to study vector fields, differential forms, and curvature tensors.

© The Author(s), under exclusive license to Springer Nature Switzerland AG 2026

N. C. Combe et al., *Exploring Information Geometry*, Frontiers in Mathematics,
https://doi.org/10.1007/978-3-032-23054-6_4

3. Global/Local Analysis

    - Differentiability connects local properties (e.g., derivatives) to global phenomena for instance using De Rham Cohomology.

4. Algebraic Structures

    - Differentiability is the gateway to advanced (and more algebraic) frameworks such as jet bundles; Lie groups (studying symmetries of differentiable manifolds); Sheaf theory (formalizing local-to-global properties in algebraic geometry).

4.1 Gateaux Directional Derivation: A Definition

Inscribing ourselves in the vein of the previous chapter, the notion of Gateaux's directional derivation appears the most natural to start with. We introduce this notion below.

Let $\mathcal{M}$ be a manifold modeled on a topological vector space $\mathcal{E}$, and let us assume that a differentiable structure can be defined on $\mathcal{E}$. As we have shown $\mathcal{E}$ to be a Hausdorff (locally convex vector space), one can define a differentiable structure via the *Gateaux directional derivation.*

4.1.1 Gateaux Derivative

Let X and Y be locally convex topological vector spaces, $U \subseteq X$ an open set, and $F : U \to Y$ a map. The *Gateaux differential* of F at $x \in U$ in the direction $\varphi \in X$ is defined as

$$\mathrm{d}F(x;\varphi) = \lim_{\substack{t\to 0\\ t\in\mathbb{K}}} \frac{F(x+t\varphi)-F(x)}{t} = \frac{\mathrm{d}}{\mathrm{d}t}F(x+t\varphi)\bigg|_{t=0} \tag{4.1}$$

provided that the limit exists in the topology of Y. Here $\mathbb{K} = \mathbb{R}$ or $\mathbb{C}$ depending on the scalar field. If this limit exists for all $\varphi \in X$, F is *Gateaux differentiable at* x.

Remark 4.1 1. **Domain Requirement**: $x + t\varphi \in U$ for all sufficiently small $|t|$ (guaranteed since U is open and the ray $\{x + t\varphi : t \in \mathbb{K}\}$ is continuous).
2. **Linearity**: Gateaux differentiability does *not* imply $\mathrm{d}F(x;\cdot)$ is linear. When $\varphi \mapsto \mathrm{d}F(x;\varphi)$ is linear and continuous, we say F has a *linear Gateaux derivative* at x.

3. **Comparison to Fréchet**:

   - Gateaux Derivative: Directional derivatives only.
   - Fréchet Derivative: Uniform approximation property.
   - Fréchet differentiability $\Rightarrow$ Gateaux differentiability (with $D_F F(x)\varphi = \mathrm{d}G(x;\varphi)$).
   - Converse fails in infinite dimensions.

4. **Weak Variant**: The *weak Gateaux derivative* replaces norm topology with weak topology: For all $\ell \in Y^*$,

$$\ell(\mathrm{d}F(x;\varphi)) = \lim_{t\to 0} \frac{\ell(F(x+t\varphi)) - \ell(F(x))}{t}.$$

4.1.2 Complex Case Specialization

When X, Y are **topological vector spaces (TVS)**, then the limit in (4.1) is taken as $t \to 0$ in $\mathbb{C}$.

- This requires complex differentiability along rays: The function $t \mapsto F(x+t\varphi)$ must be holomorphic in a neighborhood of $t = 0$.

  1. **Complex Gateaux $\Rightarrow$ real Gateaux exists.** If g is complex-differentiable at 0, then the restriction $g|_{\mathbb{R}}$ is real-differentiable. Therefore,

  $$d_{\mathbb{R}} F(x;\varphi) = g'(0)$$

  exists.
  2. Forces **Cauchy–Riemann conditions** on the map $(a,b) \mapsto F(x+(a+ib)\varphi)$.

  Indeed, decompose $t = a + ib$ with $a, b \in \mathbb{R}$, and write

  $$g(a+ib) = u(a,b) + i\, v(a,b),$$

  where $u(a,b), v(a,b) \in Y_{\mathbb{R}}$ (the realification of Y).
  The existence of the complex derivative $g'(0)$ implies:

  – The partial derivatives

  $$\frac{\partial u}{\partial a}, \quad \frac{\partial u}{\partial b}, \quad \frac{\partial v}{\partial a}, \quad \frac{\partial v}{\partial b}$$

  exist at $(a,b) = (0,0)$.

– These partial derivatives satisfy the Cauchy–Riemann equations

$$\frac{\partial u}{\partial a} = \frac{\partial v}{\partial b}, \qquad \frac{\partial u}{\partial b} = -\frac{\partial v}{\partial a},$$

at the point $(a, b) = (0, 0)$.

- **Converse fails**. Real Gâteaux differentiability does not imply complex Gâteaux differentiability in infinite dimensions.

This foundation provides the conceptual bridge to differentiation on locally convex manifolds, where directional derivatives are fundamental.

4.1.3 Differentiable Manifolds

We start with a definition on differentiable manifolds.

Definition 4.1 (Differentiable manifold) A $\mathcal{C}^k$*-differentiable manifold* $\mathcal{M}$ is a topological manifold where the condition $(\mathcal{M}3)$ is substituted by a new condition:

$(\mathcal{M}3')$ for each pair of indices $i, j \in I$, the transition map between overlapping coordinate charts (assumed non-empty),

$$\varphi_j \circ \varphi_i^{-1} : \varphi_i(\mathcal{U}_i \cap \mathcal{U}_j) \to \varphi_j(\mathcal{U}_i \cap \mathcal{U}_j),$$

is of class $\mathcal{C}^k$, meaning that it is k-times continuously differentiable. Moreover, for every $i, j \in I$, the set $\varphi_i(\mathcal{U}_i \cap U_j)$ is open in the model space $\mathcal{E}$.

Moreover, we have the following compatibility criterion.

Definition 4.2 (Compatible atlas) Two $\mathcal{C}^k$ atlases on $\mathcal{M}$, modeled on $\mathcal{E}$, are said to be compatible if their union is an another such atlas.

Remark that this notion of compatibility is in fact an equivalence relation.

Exercise 4.1 Prove the remark above.

Definition 4.3 (Admissible atlas) Given a manifold $\mathcal{M}$, all atlases, modeled on $\mathcal{E}$, and lying within the same equivalence class, are said to be *admissible* on $\mathcal{M}$.

It is enough to have an admissible atlas to define the structure of a manifold.

4.1.4 Differentiable Mappings

Definition 4.4 (Differentiable mappings) Consider a pair of $\mathcal{C}^k$-differentiable manifolds, denoted $\mathcal{M}$ and $\mathcal{M}'$.

- A mapping $f : \mathcal{M} \to \mathcal{M}'$ of class $\mathcal{C}^r$, where $r \leq k$, if for every chart $(\mathcal{U}_i, \varphi_i)$ of $\mathcal{M}$ and every chart $(\mathcal{V}_j, \psi_j)$ of $\mathcal{M}'$ such that $f(\mathcal{U}_i) \subset \mathcal{V}_j$, the mapping $\psi_j \circ f \circ \varphi_i^{-1}$ of $\varphi_i(\mathcal{U}_i)$ into $\psi_j(\mathcal{V}_j)$ is differentiable, of class $\mathcal{C}^r$.
- A *diffeomorphism* f of class $\mathcal{C}^r$ is a bijective mapping f such that f and f^{-1} are continuously $\mathcal{C}^r$-differentiable.
- If the transition maps $\varphi_j \circ \varphi_i^{-1}$ are diffeomorphisms of class $\mathcal{C}^\infty$, then the manifold $\mathcal{M}$ is said to be smooth.
- A $\mathcal{C}^k$-differentiable function $f : \mathcal{M} \to \mathbb{R}$ on $\mathcal{M}$ is a mapping of class $\mathcal{C}^k$ from $\mathcal{M}$ to $\mathbb{R}$.
- A function f is *differentiable* at a point m on the manifold $\mathcal{M}$ if, for some coordinate chart (U, ϕ) containing m, the composition $f \circ \varphi^{-1}$ is differentiable at the image point $\varphi(m)$.

Notice, that these definitions do not depend on the chart.

Remark 4.2 Since the implicit function theorem does not hold for arbitrary locally convex space, this definition has a limited utility if we do not introduce more information on the topology.

4.2 Manifolds Modeled on a Normed Vector Space

Interesting properties are obtained in the case where the model space $\mathcal{E} = \mathfrak{B}$ is a Banach vector space a natural differentiability structure is provided by the Banach derivation. In this case, the implicit function theorem allows to derive interesting properties.

Definition 4.5 (Banach manifold) A $\mathcal{C}^k$-Banach manifold $\mathcal{B}$ is a manifold such that condition $\mathcal{M}3$ is replaced by:

- $\mathcal{M}3''$. The map
$$\varphi_j \circ \varphi_i^{-1} : \varphi_i(\mathcal{U}_i \cap \mathcal{U}_j) \to \varphi_j(\mathcal{U}_i \cap \mathcal{U}_j)$$
is a $\mathcal{C}^k$-isomorphism on $\mathfrak{B}$ for each pair of indices i, j such that $\mathcal{U}_i \cap \mathcal{U}_j \neq \emptyset$ and for any $i, j \in I$ we have that $\varphi_i(\mathcal{U}_i \cap \mathcal{U}_j)$ is open in $\mathcal{B}$.

Definition 4.6 (Riemann manifold) A Riemannian manifold is a differentiable manifold modeled on a real vector space $\mathcal{E}$, the topology of which is given by a scalar product $\langle \cdot, \cdot \rangle$.

Let us recall what a scalar product is. A *scalar product* on the vector space $\mathcal{E}$ is a bilinear symmetric form

$$\langle \cdot, \cdot \rangle : \mathcal{E} \times \mathcal{E} \to \mathbb{R},$$
$$(x, y) \mapsto \langle x, y \rangle,$$

where:

1. The bilinearity property is satisfied. For any scalars α, β and for any $x, x_1, x_2, y, y_1, y_2 \in \mathcal{E}$, the following holds:

$$\langle \alpha x_1 + \beta x_2, y \rangle = \alpha \langle x_1, y \rangle + \beta \langle x_2, y \rangle$$
$$\langle x, \alpha y_1 + \beta y_2 \rangle = \alpha \langle x, y_1 \rangle + \beta \langle x, y_2 \rangle.$$

2. The scalar product is symmetric. For any $x, y \in \mathcal{E}$:

$$\langle x, y \rangle = \langle y, x \rangle.$$

3. The scalar product is positive definite. For any $x, y \in \mathcal{E}$:

$$\langle x, x \rangle \geq 0, \quad \langle x, x \rangle = 0 \iff x = 0.$$

A Riemann manifold is said to be a Banach manifold for the norm:

$$\|x\| = \langle x, x \rangle^{\frac{1}{2}}, \ x \in \mathcal{E}.$$

In the case where the model space is a Hilbert space, we speak of *Hilbert manifold*. Hence, a Riemann manifold is a real Hilbert manifold.

Remark 4.3 It is important to keep in mind that in the above definition the scalar product *must* be real-valued, due to the reliance on real analysis, positive definiteness, and geometric interpretations. Generalizations to other fields exist but define distinct structures such as in complex geometry (Hermitian manifolds use sesquilinear forms) and non-Archimedean geometry (manifolds over $\mathbb{Q}_p$ use p-adic norms but lack a metric space structure induced by an inner product).

4.2.1 Real Manifolds, Coordinates, Charts

A particularly interesting class of Riemannian manifolds is the manifold endowed with the model space $\mathcal{E} = \mathbb{R}^n$. The usual topology and differentiability structure on $\mathbb{R}^n$ induce a differentiable structure on $\mathcal{M}$, making it a real n-dimensional differentiable manifold.

In the following, by an n-dimensional (real) manifold $\mathcal{M}$, we mean a Hausdorff topological space in which every point has a neighborhood homeomorphic to $\mathbb{R}^n$.

Definition 4.7 (*n*-dimensional differentiable manifold) An n-dimensional real manifold of class C^k is a manifold modeled on $\mathbb{R}^n$ and such that the condition $\mathcal{M}3$ is replaced by the following property. For each pair of indices $i, j \in I$, the map:

$$\varphi_j \circ \varphi_i^{-1} : \varphi_i(\mathcal{U}_i \cap \mathcal{U}_j) \to \varphi_j(\mathcal{U}_i \cap \mathcal{U}_j),$$

is a $\mathcal{C}^k$-isomorphism on $\mathbb{R}^n$.

Furthermore, for any pairs of indices $i, j \in I$ the image of the intersection $\mathcal{U}_i \cap \mathcal{U}_j$ given by $\varphi_i(\mathcal{U}_i \cap \mathcal{U}_j)$, under the coordinate map φ_i, is an open subset of $\mathbb{R}^n$ (Fig. 4.1).

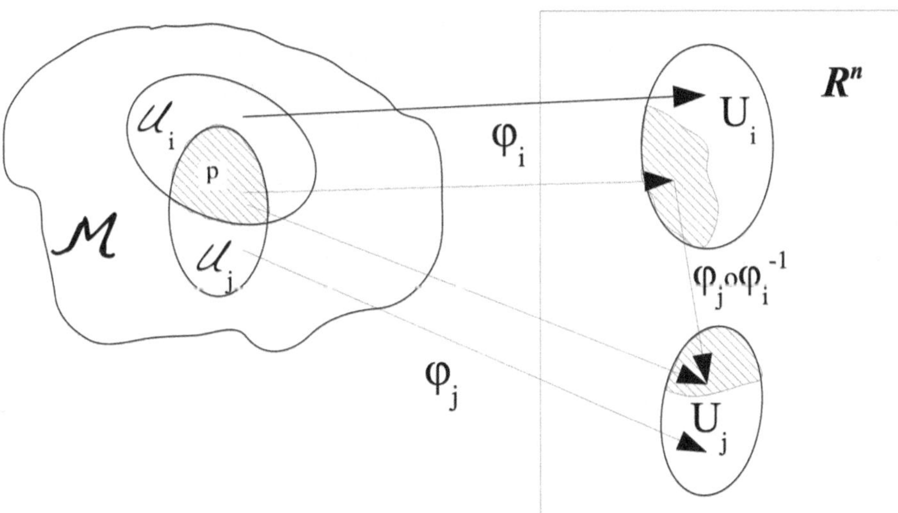

Fig. 4.1 n-dimensional real manifold

A natural way to describe the structure of an n-dimensional differentiable manifold is through local coordinate systems. These coordinate systems are given by charts, which map open subsets of the manifold to open subsets of R^n, allowing us to analyze the manifold locally as if it was a Euclidean space.

Definition 4.8 (Local coordinate system) Let $(\mathcal{U}, \varphi)$ be a chart at the point $p \in \mathcal{M}$, such that the image $\varphi(p) = x = (x^1, \ldots, x^n) \in \mathbb{R}^n$. Given a basis $\{e_1, \ldots, e_n\}$ of the Euclidean space $\mathbb{R}^n$, the coordinates $(x^1, \ldots, x^n)$ of the image $\varphi(p) \in \mathbb{R}^n$ of $p \in U \subset \mathcal{M}$ are called the *coordinates of p in the chart* $(\mathcal{U}, \varphi)$. The chart $(\mathcal{U}, \varphi)$ is also called the local coordinate system.

Local coordinate systems provide a way to describe differentiable manifolds in terms of open subsets $\mathbb{R}^n$, allowing us to define smooth functions and analyze local geometric properties. In particular, this enables the construction of tangent spaces, which capture the local linear structure of the manifold at each point. The dual spaces to these, known as cotangent spaces, naturally arise when considering differential forms and gradients of functions, playing a fundamental role in differential geometry and analysis on manifolds.

4.3 Tangent and Cotangent Spaces

To grasp the essence of the tangent and cotangent spaces, we begin with the most intuitive objects on a manifold: curves. A curve provides a way to move infinitesimally along the manifold, and by examining how functions change along such curves, we arrive at the concept of directional derivatives. These, in turn, will guide us toward a precise definition of tangent vectors. Finally, this will naturally lead us to the construction of a tangent space, the fundamental linear structure underlying the local geometry of the manifold.

4.3.1 Curves on a Manifold and Directional Derivatives

Definition 4.9 (Curve) A curve $\gamma : [a, b] \to \mathcal{M}$ on a Manifold $\mathcal{M}$ is a C^p-map, where $p \geq 1$ mapping an interval $I = [a, b] \subset \mathbb{R}$ into $\mathcal{M}$.

The curve is said to be smooth if it is a C^∞ map.

Definition 4.10 (Tangent vector to a curve) Let $\gamma : t \mapsto \gamma(t)$ be a curve of class C^1 such that:

$$\gamma(t_0) = p_0 \in \mathcal{M}$$
$$x_0 := \varphi(p_0) \in \mathbb{R}^n.$$

Then, the tangent vector u_0 at γ in p_0 is defined by

$$u_0 = \left. \frac{\mathrm{d}\gamma}{\mathrm{d}t} \right|_{t_0} . \tag{4.2}$$

Definition 4.11 (Directional derivative) Let $\mathcal{A}(p)$ be the family of C^1-functions defined on a neighborhood of $p \in \mathcal{M}$. Assume $f \in \mathcal{A}(p)$.

Consider a curve γ of class C^1,

$$\gamma : t \mapsto \gamma(t),$$

such that at some fixed parameter value t_0, we have

$$\gamma(t_0) = p_0.$$

Then, the *directional derivative* of f in the direction of the curve γ at t_0 is defined as

$$D_{u_0} f(t_0) = \left.\frac{\mathrm{d} f \circ \gamma}{\mathrm{d}t}\right|_{t_0}, \quad u_0 = \left.\frac{\mathrm{d}\gamma}{\mathrm{d}t}\right|_{t_0}, \quad \text{where} \quad f \in \mathcal{A}(p_0). \tag{4.3}$$

More generally:

Definition 4.12 (Derivation) A derivation at a point p on $\mathcal{M}$ is a linear functional

$$D : \mathcal{A}(p) \to \mathbb{R},$$

where the Leibniz rule holds:

$$D(\alpha f + \beta g)(p) = \alpha Df(p) + \beta Dg(p), \quad \alpha, \beta \in \mathbb{R}, \tag{4.4}$$

$$D(fg)(p) = [(Df)g + f Dg](p), \quad p \in \mathcal{M}. \tag{4.5}$$

4.4 Tangent Vector—Tangent Space

One can associate, via the formula (4.3), at each point $p \in \mathcal{M}$ a tangent vector X_p.

Definition 4.13 (Tangent vector) A *tangent vector* X_p at a point $p \in \mathcal{M}$ on a differentiable manifold $\mathcal{M}$ is a linear map

$$X_p : \mathcal{A}(p) \to \mathbb{R},$$

where $\mathcal{A}(p)$ denotes the space of functions defined and differentiable in some neighborhood of $p \in \mathcal{M}$.

This map is a derivation, meaning that it satisfies for any $f, g \in \mathcal{A}(p)$:

1. **Linearity**:

$$X_p(\alpha f + \beta g) = \alpha X_p(f) + \beta X_p(g) \quad \text{for} \quad \alpha, \beta \in \mathbb{R}.$$

2. **Leibniz Rule**:

$$X_p(f \cdot g) = f(p)X_p(g) + g(p)X_p(f), \tag{4.6}$$

for all $f, g \in \mathcal{A}(p)$.

Notice that two differentiable functions f_1 and f_2 which coincide on a neighborhood of p have the same tangent vector. Therefore, a tangent vector at p is the same for all functions belonging to the class of differential functions, coinciding on a neighborhood of p.

The class of functions which coincide on a neighborhood of p is called a "germ" of f. A germ forms an algebra under the sum and the point wise product. A tangent vector (also called contravariant vector) is a derivation on the algebra of germs of differentiable functions at p.

Definition 4.14 (Tangent space) The space of all tangent vectors at $p \in \mathcal{M}$, equipped with the addition and scalar multiplication defined by

$$(\alpha X_p + \beta Y_p)(f) = \alpha X_p(f) + \beta Y_p(f), \quad \alpha, \beta \in \mathbb{R}, \quad X_p, Y_p \in \mathcal{T}_p(\mathcal{M}), \tag{4.7}$$

forms a real vector space, called the *tangent space* at p, and denoted by $\mathcal{T}_p(\mathcal{M})$.

Let $\gamma \in \mathcal{A}(p)$ be a curve such that $\gamma(t_0) = p_0$ and let $u_0 = \frac{\mathrm{d}\gamma}{\mathrm{d}t}|_{t_0}$ be the tangent vector at p_0 along γ. A tangent vector X_{p_0} at $p_0 \in \mathcal{M}$ in the direction u_0 can be written

$$X_{p_0}(f) = D_{u_0} f = \left.\frac{\mathrm{d} f \circ \gamma}{\mathrm{d}t}\right|_{t_0}, \quad \forall f \in \mathcal{A}(p), \quad u_0 = \left.\frac{\mathrm{d}\gamma}{\mathrm{d}t}\right|_{t_0}. \tag{4.8}$$

In the case of an n-dimensional real manifold $\mathcal{M}$. Let $(\mathcal{U}, \varphi)$ be a chart at $p \in \mathcal{M}$ with local coordinates $x = (x^1, \dots, x^n) = \varphi(p) \in \varphi(\mathcal{U}) \subset \mathbb{R}^n$. A tangent vector at p is a derivation $X_p : C^\infty(\mathcal{M}) \to \mathbb{R}$ satisfying the Leibniz rule. Its local components $\{X_p(\varphi^i)\}_{i=1}^n$ are defined via:

$$\varphi^i = \pi^i \circ \varphi \quad \text{and} \quad \pi^i\left(x^1, \dots, x^n\right) = x^i,$$

where π^i is the ith coordinate projection in $\mathbb{R}^n$.

- The partial derivative operators $\left.\frac{\partial}{\partial x^i}\right|_p$ are derivations defined by:

$$\left.\frac{\partial}{\partial x^i}\right|_p (f) := \left.\frac{\partial(f \circ \varphi^{-1})}{\partial x^i}\right|_{x=\varphi(p)}.$$

These satisfy the Leibniz rule (and thus are valid tangent vectors).

- The set $\left\{ \frac{\partial}{\partial x^i}\Big|_p \right\}_{i=1}^n$ forms a basis for $\mathcal{T}_p(\mathcal{M})$, called the *coordinate basis* (or *natural basis*).

Remark 4.4 This basis is generally **not** orthogonal and depends on the chart $(\mathcal{U}, \varphi)$.

For any C^1-function f defined near p:

$$X_p(f) = \sum_{i=1}^n X_p(\varphi^i) \left. \frac{\partial (f \circ \varphi^{-1})}{\partial x^i} \right|_{x=\varphi(p)} . \tag{4.9}$$

Equivalently, the tangent vector decomposes as:

$$X_p = \sum_{i=1}^n X_p(\varphi^i) \left. \frac{\partial}{\partial x^i} \right|_p . \tag{4.10}$$

Remark 4.5 Any other basis (or frame) can be obtained from the local coordinate basis. Let $\{e_i\}_{i=1}^n$ be a basis of $\mathcal{T}_p(\mathcal{M})$ with:

$$e_k = \sum_{i=1}^n \Phi_k^i \frac{\partial}{\partial x^i} = \sum_{i=1}^n \Phi_k^i \partial_i ,$$

where Φ_k^i is the matrix corresponding to an invertible linear transformation. We have

$$X_p = \sum_{k=1}^n \chi_p^k e_k, \text{ with } \chi_p^k = \Phi_i^k X_p^i ,$$

the X_v^k being the component of X_p in the basis $\{e_i\}_{i=1}^n$.

Tangent and cotangent spaces are dual vector spaces, linked by a canonical pairing. We discuss now the notion of cotangent vector spaces.

4.4.1 Cotangent Vector, Cotangent Space

4.4.1.1 Cotangent Vectors and Differential 1-Forms

Definition 4.15 (Cotangent vector space) Let $\mathcal{M}$ be a manifold. The dual space $\mathcal{T}_p^\star(\mathcal{M})$ to the tangent vector space $\mathcal{T}_p(\mathcal{M})$ in $p \in \mathcal{M}$ is the space of linear forms on $\mathcal{T}_p(\mathcal{M})$. It is a vector space called the *cotangent vector space* to $\mathcal{M}$ at p.

The elements of $\mathcal{T}_p^\star(\mathcal{M})$ are called cotangent vectors, or covariant vectors, or covectors, or differential 1-forms.

Let $\omega_p \in \mathcal{T}_p^\star(\mathcal{M})$ be a differential 1-form and consider $X_p \in \mathcal{T}_p(\mathcal{M})$. Then:

$$\begin{cases} \omega_p(X_p) \in \mathbb{R}, \\ X_p(\omega_p) = \omega_p(X_p). \end{cases} \tag{4.11}$$

4.4.1.2 Finite-Dimensional Real Manifold

In the case of an n-dimensional real manifold, a useful canonical isomorphism between a space and its dual can be chosen as follows.

Let $(e_1, e_2, \ldots, e_n)$ be a basis in $\mathcal{T}_p(\mathcal{M})$. We may construct its dual $(\varepsilon^1, \varepsilon^2, \ldots, \varepsilon^n)$ by

$$\varepsilon^i(X_p) = X_p^i, \tag{4.12}$$

where X_p^i is the component of X_p in the basis $\{e_j\}$ and

$$\varepsilon^i(e_j) = \delta_j^i. \tag{4.13}$$

If we have chosen the natural basis $\{e_i = \partial/\partial x^i\}_{i=1}^n$, then the dual basis is denoted by $\{\epsilon^i = \mathrm{d}x_i\}_{i=1}^n$ with

$$\mathrm{d}x_i\left(\frac{\partial}{\partial x^j}\right) = \frac{\partial}{\partial x^j}(\mathrm{d}x_i) = \delta_j^i; \tag{4.14}$$

a tangent vector $X_p \in \mathcal{T}_p$ takes the form

$$X_p = \sum_{i=1}^n X_p^i \frac{\partial}{\partial x^i}, \quad X_p^i = \mathrm{d}x_i(X_p);$$

whereas a cotangent vector $\omega_p \in \mathcal{T}_p^\star(\mathcal{M})$ in the dual basis is given by

$$\omega_p = \sum_{i=1}^n \omega_{pi}\, \mathrm{d}x_i, \qquad \omega_{pi} = \omega\left(\frac{\partial}{\partial x^i}\right) = \frac{\partial \omega}{\partial x^i}.$$

For a real manifold of finite dimension, we have the following equality:

$$\mathcal{T}_p^{\star\star}(\mathcal{M}) = \mathcal{T}_p(\mathcal{M}). \tag{4.15}$$

4.4.2 Implications of a Chart Change

Let $\mathcal{M}$ be a n-dimensional real manifold. A chart $(\mathcal{U}, \varphi)$ with local coordinates $\varphi : p \to \varphi(p) = x = (x^1, \ldots, x^n) \in \mathbb{R}^n$ at $M \in \mathcal{M}$ induces a local coordinate basis on the tangent space given by

$$\{e_i = \partial/\partial x^i\}_{i=1}^n$$

and its dual basis is

$$\{\varepsilon^i = \mathrm{d}x^i\}_{i=1}^n.$$

Had we chosen a different chart, say $(\mathcal{U}', \varphi')$ with $\varphi' : p \to x' = (x'^1, \ldots, x'^n)$ at $p \in \mathcal{M}$, a local coordinate basis can be given by $\{e'_i = \partial/\partial x'^i\}_{i=1}^n$ with dual basis $\{\varepsilon'^i = \mathrm{d}x'^i\}_{i=1}^n$. We can certainly express one local coordinate basis in terms of the other on the open set $\varphi(\mathcal{U}) \cap \varphi'(\mathcal{U}')$.

Let us break this statement down. Under a given change of a coordinates, the local coordinates of a tangent vector are described as follows:

$$X_p = \sum_{i=1}^n X^i_p \frac{\partial}{\partial x^i} = \sum_{i=1}^n X'^i_p \frac{\partial}{\partial x'^i}, \qquad \frac{\partial}{\partial x^i} = \sum_{j=1}^n \left.\frac{\partial x'^j}{\partial x^i}\right|_{\varphi(p)} \frac{\partial}{\partial x'^j}.$$

The vector transformation law is given by:

$$X'^i_p = \sum_{j=1}^n \left.\frac{\partial x'^i}{\partial x^j}\right|_{\varphi(p)} X^j_p. \tag{4.16}$$

Similarly, for the covector transformation we have

$$\omega_p = \sum_{i=1}^n \omega_i \mathrm{d}x^i = \sum_{i=1}^n \omega'_i \mathrm{d}x'^i, \qquad \omega_i = \omega\left(\frac{\partial}{\partial x^i}\right) \; \omega'_i = \omega\left(\frac{\partial}{\partial x'^i}\right)$$

$$\begin{aligned} \mathrm{d}x^i &= \sum_{j=1}^n \left.\frac{\partial x^i}{\partial x'^j}\right|_{\varphi'(p)} \mathrm{d}x'^j \\ \omega'_i &= \sum_{j=1}^n \left.\frac{\partial x^j}{\partial x'^i}\right|_{\varphi'(p)} \omega_j. \end{aligned} \tag{4.17}$$

Having explored the implications of a changing the chart for vectors and covectors, a logical next step is to establish a rigorous definition of a function's differential.

4.4.3 Differential of a Function

Definition 4.16 Let f be a differentiable function. The differential $\mathrm{d}f|_p$ of f on $\mathcal{M}$ in the neighborhood of a point p is defined by the 1-form

$$\mathrm{d}f|_p(X_p) = X_p(f). \tag{4.18}$$

In the natural basis

$$\mathrm{d}f|_p(X_p) = \sum_i X_p^i \frac{\partial f}{\partial x^i}\bigg|_p .$$

More generally, let $\{e_i\}_{i=1}^n$ be a basis in $\mathcal{T}_p(\mathcal{M})$ and let $\{\varepsilon_i\}_{i=1}^n$ be its dual basis. The value of $\mathrm{d}f|_p \in \mathcal{T}_p^\star(\mathcal{M})$ at $X_p \in \mathcal{T}_p(\mathcal{M})$ is given by

$$\mathrm{d}f|_p(X_p) = X_p(f) = \sum_i X^i e_i(f) = \sum_i e_i(f)\varepsilon^i(X_p).$$

Hence,

$$\mathrm{d}f|_p = \sum_i e_i(f)\varepsilon^i. \tag{4.19}$$

Having previously established the fundamental geometric notions of vectors, differential forms (linear maps on vectors), along with differentials of functions (a type of 1-form), we naturally arrive to the notion of tensors, which encode multilinear maps between these objects.

4.5 Tensor at a Point of a Manifold

For simplicity, in this section, we assume that the manifold is finite dimensional.

4.5.1 Universal Property

Let $V_1, V_2, \ldots, V_k$ be vector spaces over $\mathbb{R}$. The tensor product $V_1 \otimes V_2 \otimes \cdots \otimes V_k$ is a vector space together with a multilinear map

$$\tau : V_1 \times V_2 \times \cdots \times V_k \to V_1 \otimes V_2 \otimes \cdots \otimes V_k$$

such that for every vector space W and every multilinear map

$$f : V_1 \times \cdots \times V_k \to W,$$

there exists a unique linear map

$$\tilde{f} : V_1 \otimes \cdots \otimes V_k \to W$$

satisfying

$$f = \tilde{f} \circ \tau.$$

4.5.2 Multilinear Maps and Tensors of Type (r, s)

Definition 4.17 Given a point $p \in \mathcal{M}$ in an n-dimensional manifold $\mathcal{M}$, a tensor $T_p^{(r,s)}$ of type (r, s) is a multilinear form:

$$T_p : \underbrace{\mathcal{T}_p^\star \times \cdots \times \mathcal{T}_p^\star}_{r} \times \underbrace{\mathcal{T}_p \times \cdots \times \mathcal{T}_p}_{s} \to \mathbb{R},$$

$$\underbrace{(\omega_1, \ldots, \omega_r}_{r \text{ covectors}}, \underbrace{X_1, \ldots, X_s)}_{s \text{ vectors}} \mapsto T_p(\omega_1, \ldots, \omega_r, X_1, \ldots, X_s).$$

Furthermore, this object is multilinear:

$$\begin{aligned} T_p(\omega_1, \ldots, \alpha\omega_j + \beta\eta_j, \ldots, \omega_r, X_1, \ldots, X_s) \\ = \alpha T_p(\omega_1, \ldots, \omega_j, \ldots, \omega_r, X_1, \ldots, X_s) \\ + \beta T_p(\omega_1, \ldots, \eta_j, \ldots, \omega_r, X_1, \ldots, X_s), \end{aligned}$$

$$\begin{aligned} T_p(\omega_1, \ldots, \omega_r, X_1, \ldots, \alpha X_k + \beta Y_k, \ldots, X_s) \\ = \alpha T_p(\omega_1, \ldots, \omega_r, X_1, \ldots, X_k, \ldots, X_s) \\ + \beta T_p(\omega_1, \ldots, \omega_r, X_1, \ldots, Y_k, \ldots, X_s), \end{aligned}$$

for scalars $\alpha, \beta \in \mathbb{R}$, all $1 \le j \le r$, and all $1 \le k \le s$.

The following denominations are often used in the literature:

- A **contravariant tensor of order** r at M is a tensor on $\mathcal{T}_p^{\otimes r}$ (type $(r, 0)$).
- A **covariant tensor of order** s at M is a tensor on $(\mathcal{T}_p^\star)^{\otimes s}$ (type $(0, s)$).

4.5.2.1 Tensor Product and Its Universal Property

An (r, s)-tensor at a point p of a manifold is defined as a multilinear map

$$T : (\mathcal{T}_p^*(\mathcal{M}))^r \times (\mathcal{T}_p(\mathcal{M}))^s \to \mathbb{R}.$$

By the universal property, this is equivalent to a linear map

$$\widetilde{T} : (\mathcal{T}_p^*(\mathcal{M}))^{\otimes r} \otimes (\mathcal{T}_p(\mathcal{M}))^{\otimes s} \to \mathbb{R}.$$

But linear functionals on a vector space U are the same as elements of the dual space U^*. Hence

$$T \in \left((\mathcal{T}_p^*(\mathcal{M}))^{\otimes r} \otimes (\mathcal{T}_p(\mathcal{M}))^{\otimes s}\right)^*.$$

In finite dimensions, we have the canonical isomorphism

$$(U^* \otimes V^*) \cong (U \otimes V)^*,$$

and also

$$(V^*)^* \cong V.$$

To conclude, the space of tensors of type (r, s) can be identified with the tensorial product space:

$$\mathcal{T}_p(\mathcal{M})^{\otimes r} \otimes \mathcal{T}_p^\star(\mathcal{M})^{\otimes s}.$$

4.5.3 Tensor Products

The space of such tensors is isomorphic to $(\mathcal{T}_p)^{\otimes r} \otimes (\mathcal{T}_p^\star)^{\otimes s}$ and has dimension n^{r+s}. If T_p and S_p are tensors of type (r, s), then

$$(\alpha T_p + \beta S_p)(\omega_1, \ldots, \omega_r, X_1, \ldots, X_s) = \alpha T_p(\cdots) + \beta S_p(\cdots).$$

Definition 4.18 (Tensor product) Let T_p be a tensor at the point $p \in \mathcal{M}$ of type (r, s) and let S_p be a tensor of type (p, q). Then, the tensor $T_p \otimes S_p$ defined by

$$\begin{aligned} &T_p \otimes S_p(\omega_1, \ldots, \omega_r, \ldots, \omega_{r+p}, X_1, \ldots, X_s, \ldots, X_{s+q}) \\ &\quad = T_p(\omega_1, \ldots, \omega_r, X_1, \ldots, X_s) \times S_p(\omega_{r+1}, \ldots, \omega_{r+p}, \ldots, X_{s+1}, \ldots, X_{s+q}) \end{aligned} \tag{4.20}$$

is called the tensorial product of T_p and S_p.

Admitting this definition, an (r, s)-tensor T can be seen as the tensorial product of r tangent vectors and s covectors at $M \in \mathcal{M}$ as follows:

$$\begin{aligned} T(\omega_1, \ldots, \omega_r, X_1, \ldots, X_s) = &\left(T^{(1)} \otimes \cdots \otimes T^{(r)} \otimes T_{(1)} \otimes \cdots \otimes T_{(s)}\right) \\ &(\omega_1, \ldots, \omega_r, X_1, \ldots, X_s), \end{aligned}$$

where the $T^{(i)}$ are tangent vectors (contravariant) and the $T_{(k)}$ are covectors (covariant).

By abuse of notation, we have omitted the index p to have a more convenient formula. However, it is important to remember that the definition is local.

Let us choose a basis of $\mathcal{T}_p(\mathcal{M})^{\otimes r} \otimes \mathcal{T}_p^\star(\mathcal{M})^{\otimes s}$ given by the n^{r+s} elements:

$$\mathbf{e}_{i_1 \cdots i_r}^{j_1 \cdots j_s} = e_{i_1} \otimes \cdots \otimes e_{i_r} \otimes \varepsilon^{j_1} \otimes \cdots \otimes \varepsilon^{j_s}, \tag{4.21}$$

where $\{e_i\}_{i=1}^n$ is a basis for $\mathcal{T}_p(\mathcal{M})$, and $\{\varepsilon^j\}_{j=1}^n$ is its dual basis for $\mathcal{T}_p^\star(\mathcal{M})$ satisfying $\varepsilon^j(e_i) = \delta_i^j$.

If we work in the framework of local coordinate basis, the associated basis of the (r, s)-tensor space is

$$\mathbf{e}_{i_1\cdots i_r}^{j_1\cdots j_s} = \partial_{i_1} \otimes \cdots \otimes \partial_{i_r} \otimes \mathrm{d}x^{j_1} \otimes \cdots \otimes \mathrm{d}x^{j_s} = \bigotimes_{k=1}^{r} \partial_{i_k} \bigotimes_{\ell=1}^{s} \mathrm{d}x^{j_\ell}, \quad 1 \le i_k, j_\ell \le n,$$

where $\{\partial_i\}_{i=1}^n$ is the coordinate basis for $\mathcal{T}_p(\mathcal{M})$ and $\{\mathrm{d}x^j\}_{j=1}^n$ is its dual basis for $\mathcal{T}_p^\star(\mathcal{M})$ satisfying $\mathrm{d}x^j(\partial_i) = \delta_i^j$.

Then, the (r, s)-tensor T is given by:

$$T = \sum_{\substack{i_1,\ldots,i_r=1 \\ j_1,\ldots,j_s=1}}^{n} T_{j_1\cdots j_s}^{i_1\cdots i_r} \mathbf{e}_{i_1\cdots i_r}^{j_1\cdots j_s}. \tag{4.22}$$

To enhance the clarity, it is advantageous to adopt the Einstein summation convention, where repeated upper and lower indices are implicitly summed over.

In the following formula, we illustrate this convention with the expansion of a tensor T:

$$T = T^{i_1\cdots i_r}{}_{j_1\cdots j_s} \partial_{i_1} \otimes \cdots \otimes \partial_{i_r} \otimes \mathrm{d}x^{j_1} \otimes \cdots \otimes \mathrm{d}x^{j_s}, \tag{4.23}$$

where the components are given by:

$$T_{i_1\cdots i_r j_1\cdots j_s}^{i_1\cdots i_r} = T(\mathrm{d}x^{i_1}, \ldots, \mathrm{d}x^{i_r}, \partial_{j_1}, \ldots, \partial_{j_s}), \tag{4.24}$$

Under a change of coordinates $\{x^i\} \to \{x'^i\}$, an (r, s)-tensor transforms via r applications of the Jacobian (for contravariant indices) and s applications of the inverse Jacobian (for covariant indices):

$$T'^{i'_1\cdots i'_r}{}_{j'_1\cdots j'_s} = \prod_{k=1}^{r} \frac{\partial x'^{i'_k}}{\partial x^{i_k}} \prod_{\ell=1}^{s} \frac{\partial x^{j_\ell}}{\partial x'^{j'_\ell}} T^{i_1\cdots i_r}{}_{j_1\cdots j_s}. \tag{4.25}$$

We leverage this framework to study symmetries of tensors. The study of tensor symmetries within this framework reveals fundamental insights into their invariant properties and multilinear algebraic behavior.

4.5.4 Symmetry Properties of Tensors

Let us first consider a covariant tensor of order r (i.e., a $(0, r)$-tensor). Let S_r be the group of permutations of $\{1, 2, \ldots, r\}$. It acts on the tensor at $M \in \mathcal{M}$ via:

$$(\sigma T)(X_1, \ldots, X_r) = T(X_{\sigma^{-1}(1)}, \ldots, X_{\sigma^{-1}(r)}), \quad \sigma \in S_r, \ X_k \in \mathcal{T}_p(\mathcal{M}),$$

with corresponding component transformation:

$$(\sigma T)_{i_1 \cdots i_r} = T_{i_{\sigma(1)} \cdots i_{\sigma(r)}},$$

where T has the symmetry defined by σ.

Definition 4.19 • If $\sigma T = T$ for all $\sigma \in S_r$, then T is **symmetric**.
• If $\sigma T = \mathrm{sign}(\sigma) \cdot T$ for all $\sigma \in S_r$, then T is **antisymmetric**.

The symmetrization operator Sym and antisymmetrization operator Alt are defined as:

$$\begin{aligned} \mathrm{Sym}\, T &= \frac{1}{r!} \sum_{\sigma \in S_r} \sigma T && \text{(completely symmetric tensor)} \\ \mathrm{Alt}\, T &= \frac{1}{r!} \sum_{\sigma \in S_r} \mathrm{sign}(\sigma) \cdot \sigma T && \text{(completely antisymmetric tensor).} \end{aligned} \tag{4.26}$$

In local coordinates, antisymmetrization is expressed as:

$$(\mathrm{Alt}\, T)_{i_1 \cdots i_r} = \frac{1}{r!} \delta^{k_1 \cdots k_r}_{i_1 \cdots i_r} T_{k_1 \cdots k_r} \tag{4.27}$$

using Einstein summation and the generalized Kronecker tensor:

$$\delta^{k_1 \cdots k_r}_{i_1 \cdots i_r} = \begin{cases} \mathrm{sign}(\sigma) & \text{if } (k_1, \ldots, k_r) = (i_{\sigma(1)}, \ldots, i_{\sigma(r)}) \text{ for some } \sigma \in S_r \\ 0 & \text{otherwise} \end{cases}.$$

The symmetry properties for contravariant tensors are defined analogously.

Remark 4.6 1. Permutation symmetry applies only to indices of the same type (covariant or contravariant).
2. Square brackets $[\cdot]$ denote antisymmetrization and parentheses $(\cdot)$ denote symmetrization.
3. Example: For a (3, 2)-tensor, $T^{[ab]c}_{(de)}$ is antisymmetric in a, b (contravariant indices) and symmetric in d, e (covariant indices).

5 Fiber Bundles

In the language of modern geometry, fiber bundles serve as a unifying framework, extending classical notions such as vector bundles to a broader setting. They naturally arise in various mathematical and physical contexts, providing a structural backbone for spaces with local product structures.

A principal application appears in Gauge Theory, where principal bundles furnish a natural setting for describing gauge fields in physics, including the Yang–Mills theories. Here, the connection on a bundle encodes the dynamics of fundamental interactions, and curvature expresses field strength in a geometrically intrinsic way.

In topology, fiber bundles play a fundamental role in the theory of characteristic classes, such as the Chern classes, which assign global topological invariants to vector bundles. These invariants capture deep global properties of manifolds, serving as indispensable tools in the study of topology, geometry, and mathematical physics.

5.1 Definitions and Examples

The notion of bundle have been introduced to generalize topological product.

Definition 5.1 (Fiber bundle) A fiber bundle is a triple $(\mathcal{B}, \mathcal{M}, \pi)$ consisting of two topological spaces $\mathcal{B}$ and $\mathcal{M}$ and a continuous surjective map

$$\pi : \mathcal{B} \to \mathcal{M},$$

where π is called the projection of the total space $\mathcal{B}$ onto the base space $\mathcal{M}$.

© The Author(s), under exclusive license to Springer Nature Switzerland AG 2026

N. C. Combe et al., *Exploring Information Geometry*, Frontiers in Mathematics,
https://doi.org/10.1007/978-3-032-23054-6_5

The topological space $\mathcal{F}_p = \pi^{-1}(p),\ p \in \mathcal{M}$ is called the fiber at $p \in \mathcal{M}$.

Example 5.1 The simplest example of bundle is the product bundle defined by $(E_1 \times E_2, E_1, \pi)$ where E_i is isomorphic to $\mathbb{R}$ and with $\pi(x, y) = x$, for all $x \in E_1$ and $y \in E_2$. This is illustrated by the following diagram.

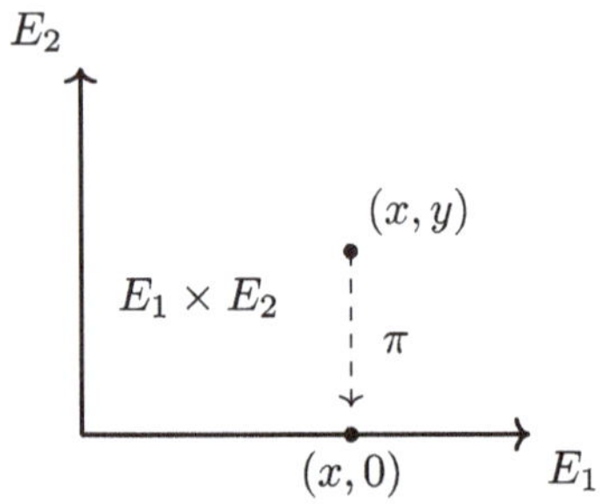

So, in this example the total space is given by the topological product $\mathcal{B} = E_1 \times E_2$ and the base space is $\mathcal{M} = E_1$. The projection map is given by $\pi(x, y) = x$, where we project $E_1 \times E_2$ onto E_1. Naturally, we could have chosen to project on E_2.

The need to generalize topological products can be seen on the following examples.

(a) The cylinder is obtained by taking the topological product $S^1 \times I$, where I is a segment. The fiber bundle is defined by the triple $(S^1 \times I, S^1, \pi)$, where here we have chosen to define the projection map as follows $\pi : S^1 \times I \to S^1$.

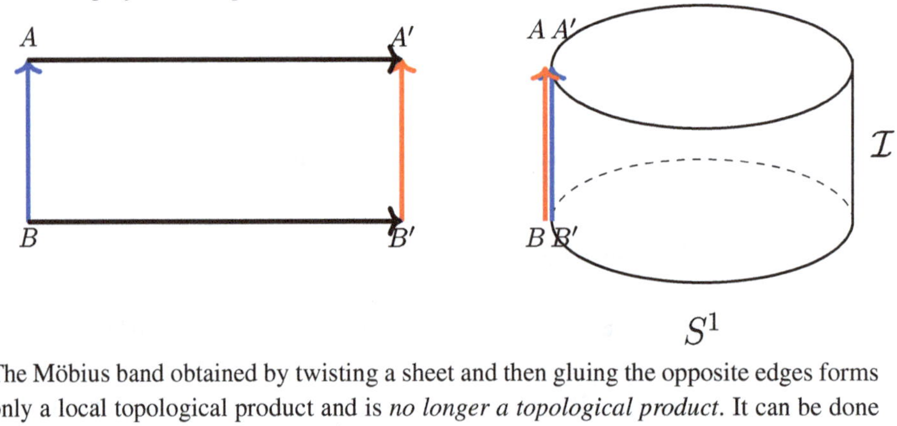

(b) The Möbius band obtained by twisting a sheet and then gluing the opposite edges forms only a local topological product and is *no longer a topological product*. It can be done locally: for an open subset $U \subset S^1$, the topological product $U \times I$ describes a segment of the Möbius band, but does not take under account the twisting operation.

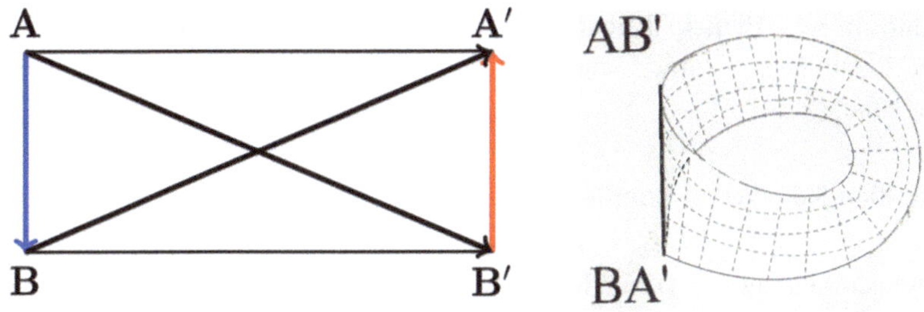

5.1.1 Properties

In the following we restrict ourselves to the case where the topological space $\mathcal{F}_p = \pi^{-1}(p),\ p \in \mathcal{M}$ are homeomorphic to a space F, called typical fiber.

Definition 5.2 (Trivial bundle) A trivial bundle $\mathcal{B}$ is a fiber bundle homeomorphic to a product bundle $\mathcal{M} \times \mathcal{F}$ and π is just the projection from the product space to $\mathcal{M}$.

Definition 5.3 (Vector bundle) If $\mathcal{F}_M$ is a vector space, for all $M \in \mathcal{M}$ then $(\mathcal{B}, \mathcal{M}, \pi)$ is called vector bundle.

Definition 5.4 (G-bundle) A fiber bundle $(\mathcal{B}, \mathcal{M}, \pi, G)$ is made of:

1. a bundle $(\mathcal{B}, \mathcal{M}, \pi)$ together with a typical fiber $\mathcal{F}$;
2. a topological group G of homeomorphisms of $\mathcal{F}$ onto itself (called the structural group);
3. a covering by open sets $\{U_j\}_{j \in J \subseteq \mathbb{N}}$ where $U_j \in \mathcal{M}$ such that:

   a. Locally, the bundle is a trivial bundle.
   b. Let $p \in U_i \cap U_j$, the homeomorphism $\psi_{M_i} \circ \psi_{M_j}^{-1} : \mathcal{F} \to \mathcal{F}$, where ψ_{M_i} denote a homeomorphism from $\mathcal{F}_x$ onto $\mathcal{F}$, is an element of the structural group G for all $i, j \in J$.
   c. The induced mapping $\gamma_{ij} : U_i \cap U_j \to G$ by $\gamma_{ij}(x) = \psi_{M_i} \circ \psi_{M_j}^{-1}$ is continuous.

Definition 5.5 (Principal fiber bundle) A G-bundle $(\mathcal{B}, \mathcal{M}, \pi, G)$ in which the typical fiber $\mathcal{F}$ and the structural group G are isomorphic by left translation[1] is called a principal fiber bundle.

Definition 5.6 (Vector bundle) If $\mathcal{F}$ is a vector space, $(\mathcal{B}, \mathcal{M}, \pi)$ is called vector bundle.

A vector bundle is a G-bundle with G the linear group of $\mathcal{F}$.

Definition 5.7 (Cross-section of a bundle) A cross-section (or section) of the bundle $(\mathcal{B}, \mathcal{M}, \pi, G)$ is a mapping $\sigma : \mathcal{M} \to \mathcal{B}$ such that $\pi \circ \sigma = I_{\mathcal{M}}$ (the identity on $\mathcal{M}$).

Theorem 5.1 *A principal fiber bundle* $(\mathcal{B}, \mathcal{M}, \pi, G)$ *is trivial if and only if it admits a global continuous cross-section.*

Proof $(\Rightarrow)$ Suppose the bundle is trivial, i.e., there exists a G-equivariant diffeomorphism $\Phi : \mathcal{B} \to \mathcal{M} \times G$ commuting with projections. Define $\sigma : \mathcal{M} \to \mathcal{B}$ by $\sigma(x) = \Phi^{-1}(x, e)$ where $e \in G$ is the identity. Then $\pi(\sigma(x)) = \pi(\Phi^{-1}(x, e)) = x$ since Φ preserves fibers. Thus σ is a global continuous cross-section.
$(\Leftarrow)$ Suppose there exists a continuous cross-section $\sigma : \mathcal{M} \to \mathcal{B}$ with $\pi \circ \sigma = \mathrm{id}_{\mathcal{M}}$. For any $p \in \mathcal{B}$, let $x = \pi(p)$. By the free transitive action of G on fibers, there exists a unique $g_p \in G$ such that $p = g_p \cdot \sigma(x)$. Define $\Phi : \mathcal{B} \to \mathcal{M} \times G$ by $\Phi(p) = (x, g_p)$. This map is:

- *Continuous*: As σ is continuous and the group action is continuous.
- *G-equivariant*: $\Phi(g' \cdot p) = (x, g'g_p) = g' \cdot \Phi(p)$.
- *Fiber-preserving*: $\pi_1(\Phi(p)) = x = \pi(p)$.
- *Bijective*: Inverse is $\Phi^{-1}(x, g) = g \cdot \sigma(x)$.

Thus Φ is a trivialization, making $\mathcal{B}$ trivial.

5.1.2 Tangent Bundles and Frame Bundles

The notion of tangent spaces can be upgraded to the structure of a tangent bundle. We describe this in the next section below.

[1] Let G be a group the action

$$L_g : G \to G \text{ by } L_g(h) = gh$$

is called left translation.

5.1.3 Tangent Bundle

Definition 5.8 (Tangent bundle) A tangent bundle $(\mathcal{T}(\mathcal{M}), \mathcal{M}, \pi, G)$ is a bundle with fibers being the tangent spaces.

In the case of a n-dimensional manifold, the tangent bundle $(\mathcal{T}(\mathcal{M}), \mathcal{M}, \pi, G)$ can be identified to the natural bundle defined by

- a fiber at $p \in \mathcal{M}$, $\mathcal{F}_p = \mathcal{T}_p(\mathcal{M})$ and typical fiber $\mathbb{R}^n$;
- a projection $\pi : (p, X_p) \mapsto p$;
- the structural group $G = GL(n, \mathbb{R})$ of linear automorphisms on $\mathbb{R}^n$.

5.1.4 Frame Bundle

Let $\mathcal{M}$ be an n-dimensional manifold. A **frame** at $p \in \mathcal{M}$ is an ordered basis $\phi_M = \{f_1, \dots, f_n\}$ for the tangent space $\mathcal{T}_p(\mathcal{M})$. Given any basis $\{e_1, \dots, e_n\}$ of $\mathcal{T}_p(\mathcal{M})$, there exists a unique invertible matrix $A = (a_i^j) \in GL(n, \mathbb{R})$ such that

$$f_i = a_i^j e_j \quad \text{for} \quad i \in \{1, \dots, n\},$$

using the Einstein summation convention. This establishes a bijection between the set of frames at p and $GL(n, \mathbb{R})$.

Definition 5.9 (Frame bundle) The **frame bundle** $\Phi(\mathcal{M})$ is the disjoint union of all frames over $\mathcal{M}$:

$$\Phi(\mathcal{M}) = \coprod_{p \in \mathcal{M}} \{\text{frames at } p\}.$$

Equipped with the projection $\pi : \Phi(\mathcal{M}) \to \mathcal{M}$ defined by $\pi(\phi_M) = p$, it forms a principal fiber bundle with:

- **Total Space**: $\Phi(\mathcal{M})$ (smooth manifold structure induced from $\mathcal{M}$).
- **Base Space**: $\mathcal{M}$.
- **Structure Group**: $GL(n, \mathbb{R})$ acting freely and transitively on fibers via

$$R_g(\phi_M) = \left\{ \sum_{k=1}^n g_k^1 f_k, \dots, \sum_{k=1}^n g_k^n f_k \right\}$$

  for $g = (g_i^j) \in GL(n, \mathbb{R})$.
- **Local Trivialization**: For $U \subset \mathcal{M}$ open, $\pi^{-1}(U) \cong U \times GL(n, \mathbb{R})$.

Thus $\Phi(\mathcal{M})$ is denoted as $(\Phi(\mathcal{M}), \mathcal{M}, \pi, GL(n, \mathbb{R}))$.

Furthermore, the frame bundle associated with a rank-n vector bundle E forms a principal $GL(n, \mathbb{K})$-bundle, where $\mathbb{K}$ is $\mathbb{R}$ or $\mathbb{C}$. The general linear group acts freely and transitively on frames via basis changes:

$$R_g(\phi_p) = \left\{ \sum_{k=1}^{n} g_k^1 f_k, \ldots, \sum_{k=1}^{n} g_k^n f_k \right\}, \quad g = (g_i^j) \in GL(n, \mathbb{K})$$

The topology on the frame bundle is constructed using local trivializations of E. For each local trivialization $\psi_i : \pi_E^{-1}(U_i) \to U_i \times \mathbb{K}^n$ of E, there is an induced local trivialization of the frame bundle:

$$\widetilde{\psi}_i : \pi^{-1}(U_i) \to U_i \times GL(n, \mathbb{K})$$

defined by sending a frame $\phi_p = \{f_1, \ldots, f_n\}$ at $p \in U_i$ to (p, g), where $g \in GL(n, \mathbb{K})$ is the transition matrix from ψ_i-induced basis to ϕ_p. The bundle topology is generated by these trivializations, ensuring compatibility via transition functions $g_{ij}(p) = \psi_i \circ \psi_j^{-1}|_p$.

The fibers $\pi^{-1}(p)$ are $GL(n, \mathbb{K})$-torsors. A *torsor* (principal homogeneous space) for a Lie group G is a set X with a free and transitive G-action, i.e., for any $x, y \in X$, there exists a unique $g \in G$ such that $y = g \cdot x$.

Exercise 5.1 Prove that when $\mathcal{M}$ is equipped with a Riemannian metric the structure group reduces to the orthogonal group $O(n)$.

Exercise 5.2 Prove that when $\mathcal{M}$ is a $2n$-dimensional symplectic manifold then it has a natural $Sp(2n, \mathbb{R})$-structure.

We remark the following fact, relating the manifold structure and the frame bundle.

A natural manifold structure can be given if we notice that the open sets of the typical fiber $GL(n, \mathbb{R})$ are in bijection with the open sets of $\mathbb{R}^{n^2}$, and that the structural group of diffeomorphisms of the typical fiber onto itself is simply $GL(n, \mathbb{R})$.

5.2 Vector Fields, Tangent Bundles, Lie Algebras

Given a differentiable manifold $\mathcal{M}$, we consider at each point $\hat{p}$ of $\mathcal{M}$ its corresponding tangent space $\mathcal{T}_{\hat{p}}(\mathcal{M})$. A vector field X assigns to every point $\hat{p} \in \mathcal{M}$ a tangent vector $X_{\hat{p}} \in \mathcal{T}_{\hat{p}}(\mathcal{M})$ in a smooth manner. This means that if f is a smooth function on $\mathcal{M}$, the map $\hat{p} \mapsto X_{\hat{p}}(f)$ must also be smooth.

In more concrete words, given a subset of the Euclidean space $\mathbb{R}^n$, a vector field is represented by a vector-valued function.

More fundamentally, a smooth vector field X is a linear map

$$X : C^\infty(\mathcal{M}) \to C^\infty(\mathcal{M}),$$

where $C^\infty(\mathcal{M})$ is the algebra of smooth functions and where X satisfies the Leibniz rule. This property reveals the deep algebraic nature of vector fields: they form a Lie algebra under the Lie bracket, defined as

$$[X, Y](f) = X(Y(f)) - Y(X(f)),$$

with $f \in C^\infty(\mathcal{M})$ and X, Y smooth vector fields on $\mathcal{M}$.

To link with the previous section on tangent bundles, let us highlight that a vector field on $\mathcal{M}$ can be defined in terms of a section of the tangent bundle. This is stated more precisely in the following definition.

Definition 5.10 (Vector field) A vector field X on a manifold $\mathcal{M}$ is a cross-section of the tangent bundle $\mathcal{T}(\mathcal{M})$.

Using this approach, we can say that a vector field X associates to each $\hat{p} \in \mathcal{M}$ a tangent vector $X_{\hat{p}} \in \mathcal{T}_{\hat{p}}(\mathcal{M})$ by the mapping

$$X : \hat{p} \mapsto (\hat{p}, X_{\hat{p}}).$$

A vector field $X \in \mathcal{T}(\mathcal{M})$ acts on a differentiable function f on $\mathcal{M}$ by

$$(Xf)(\hat{p}) = X_{\hat{p}}(f), \tag{5.1}$$

or in local coordinates,

$$X = \sum_{i=1}^{n} X^i \frac{\partial}{\partial x^i}, \quad X^i = X(\varphi^i), \tag{5.2}$$

where X^i are functions such that in a chart $(\mathcal{U}, \varphi)$ in the neighborhood of $\mathcal{M}$, the components are given by $X^i_{\hat{p}} = (X(\varphi^i))(\hat{p})$. These are the components of X with respect to the local coordinates $x^i = \varphi^i(\hat{p})$.

Definition 5.11 (C^r-vector field) A vector field on a C^k-manifold $\mathcal{M}$ is C^r-differentiable if the mapping $\mathcal{M} \to \mathcal{T}(\mathcal{M})$ is C^r-differentiable, where $r \leq k - 1$.

Returning to our first definition, we can refine it: for $\mathcal{C}^k$-differentiable functions, a vector field X on a $\mathcal{C}^k$-manifold $\mathcal{M}$ can be understood as a derivation on the algebra $\mathcal{C}^k(\mathcal{M})$ of $\mathcal{C}^k$ functions on $\mathcal{M}$:

$$X : \mathcal{C}^k(\mathcal{M}) \to \mathcal{C}^k(\mathcal{M}),$$
$$X(f)(\hat{p}) = (Xf)(\hat{p}), \quad \text{denoted simply } X(f) = Xf.$$

5.2.1 Vector Field Lie Algebra

Let us introduce the space $\mathfrak{T}(\mathcal{M})$ of all $\mathcal{C}^\infty$ vectors fields on $\mathcal{M}$, that is the space of all $X(f)$ which are differentiable for all $\mathcal{C}^\infty$ functions f on $\mathcal{M}$. Under the addition and the multiplication operation:

$$(X+Y)(f) = X(f) + Y(f), \quad X, Y \in \mathfrak{T}(\mathcal{M}),\ f \in \mathcal{C}^\infty(\mathcal{M}),$$
$$gX(f) = g\big(X(f)\big), X \in \mathfrak{T}(\mathcal{M}),\ f, g \in \mathcal{C}^\infty(\mathcal{M})$$

this forms a module on the ring[2] $\mathcal{C}^\infty(\mathcal{M})$, but cannot be an algebra for the product of vector fields.

The product XY of two vector fields defined by $(XY)(f) = X(Y(f))$ does not satisfy the Leibnitz rule. Indeed,

$$\begin{aligned}(XY)(fg) &= X(Y(fg)) = X\big(fY(g) + gY(f)\big)\\ &= X(f)Y(g) + fX(Y(g)) + X(g)Y(f) + gX(Y(f))\\ &\neq f(XY(g)) + g(XY(f)) = g(X(Y(f)) + f(X(Y(g)).\end{aligned}$$

Therefore XY does not form a vector field.

However, notice that the Lie bracket $[\cdot, \cdot]$ defined by

$$[X, Y]f = X(Y(f)) - Y(X(f)), \quad X, Y \in \mathfrak{T}, \tag{5.3}$$

does form a vector field. The multiplication, defined by the lie bracket is:

---

[2] A ring is a set $\mathfrak{R}$ with two internal laws, addition and multiplication. For the addition, $\mathfrak{R}$ is an abelian group, the multiplication is associative and distributive with respect to addition.

$$(xy)z = x(yz), \quad x(y+z) = xy + xz, \quad (y+z)x = yx + zx, \quad \forall x, y, z \in \mathfrak{T}.$$

A module $\mathfrak{T}$ over the ring $\mathfrak{R}$ is an abelian group together with an external operation, called scalar multiplication such that

$$\alpha(x+y) = \alpha x + \alpha y, \quad (\alpha+\beta)x = \alpha x + \beta x, \quad (\alpha\beta)x = \alpha(\beta x), \quad x, y \in \mathfrak{T},\ \alpha, \beta \in \mathfrak{R}.$$

- distributive with respect to the addition;
- anti-commutative;
- not associative but satisfies the Jacobi identity:

$$[[X, Y], Z] + [[Y, Z], X] + [[Z, X], Y] = 0, \quad X, Y, Z \in \mathfrak{X}, \tag{5.4}$$

Lemma 5.1 *The set $\mathfrak{X}(\mathcal{M})$ is a Lie algebra*[3] *for the Lie bracket.*

Proof The proof is mostly described above.

Definition 5.12 (Moving frame) A set of n linearly independent differentiable vector fields $\{e_i\}_{i=1}^n$ which form a basis of the module $\mathfrak{X}(U)$ where $U \subset \mathcal{M}$ is called a moving frame.

Notice that a moving frame may not exist *globally* on $\mathcal{T}(\mathcal{M})$.

To end this section, let us mention that vector fields can be used for instance in Index Theory. The Poincaré–Hopf theorem relates zeros of vector fields on $\mathcal{M}$ to the Euler characteristic of $\mathcal{M}$.

5.3 Cotangent Bundle—Covector Field

The topic considered in this section relates to cotangent bundles. A cotangent bundle is the natural dual to the tangent bundle, encoding the differential structures of a manifold in their most intrinsic form. While the tangent bundle $\mathcal{T}(\mathcal{M})$ describes directions of motion, the cotangent bundle $\mathcal{T}^*(\mathcal{M})$ is the realm of differentials.

For a differentiable manifold $\mathcal{M}$, the cotangent space at each point consists of linear functionals on the tangent space. That is, if $X_{\hat{p}}$ is a tangent vector, an element of $\mathcal{T}^*_{\hat{p}}(\mathcal{M})$ assigns to it a real number by evaluating a differential $df_{\hat{p}}(X_{\hat{p}})$.

5.3.1 Cotangent Bundle

Definition 5.13 (Cotangent bundle) The **cotangent bundle** of a manifold $\mathcal{M}$ is the triple $(\mathcal{T}^*(\mathcal{M}), \mathcal{M}, \pi)$ where:

- $\mathcal{T}^*(\mathcal{M}) = \coprod_{\hat{p}\in\mathcal{M}} \mathcal{T}^*_{\hat{p}}(\mathcal{M})$ is the disjoint union of all cotangent spaces;
- $\pi : \mathcal{T}^*(\mathcal{M}) \to \mathcal{M}$ is the projection map defined by $\pi(\hat{p}, \omega_{\hat{p}}) = \hat{p}$.

[3] A Lie algebra is a module with bracket as internal product.

This is a vector bundle with fiber $\mathcal{T}^*_{\hat{p}}(\mathcal{M}) \cong (\mathbb{R}^n)^*$ at each $\hat{p} \in \mathcal{M}$.

In the case of an n-dimensional manifold, the cotangent bundle $(\mathcal{T}^\star(\mathcal{M}), \mathcal{M}, \pi, G)$ can be identified with the natural bundle:

- fiber at $\hat{p} \in \mathcal{M}$: $\mathcal{F}_{\hat{p}} = \mathcal{T}^\star_{\hat{p}}(\mathcal{M})$, with typical fiber $\mathcal{F} = \mathbb{R}^n$;
- projection $\pi : (\hat{p}, \omega_{\hat{p}}) \mapsto \hat{p}$;
- structural group $G = GL(n, \mathbb{R})$, the group of linear automorphisms of $\mathbb{R}^n$.

5.3.2 Covector Field

Definition 5.14 (Covector field) A $\mathcal{C}^k$-differentiable 1-form ω is a $\mathcal{C}^k$ cross-section of the cotangent bundle. It is often called a covariant vector field.

A covariant vector field ω associates to each $\hat{p} \in \mathcal{M}$ a covariant vector $\omega_{\hat{p}} \in \mathcal{T}^\star_{\hat{p}}(\mathcal{M})$ by the mapping

$$\omega : \hat{p} \mapsto (\hat{p}, \omega_{\hat{p}}).$$

The covector field $\omega \in \mathcal{T}^\star(\mathcal{M})$ acts on the vector field $X \in \mathcal{T}(\mathcal{M})$ by

$$\omega(X)(\hat{p}) \;=\; \omega_{\hat{p}}(X_{\hat{p}}). \tag{5.5}$$

If we denote by $\{dx^i\}_{i=1}^n$ the dual basis of the natural basis $\{\partial_i = \partial/\partial x^i\}_{i=1}^n$, we have

$$dx^i(\partial_j) = \delta^i_j,$$

and

$$\omega_{\hat{p}}(X_{\hat{p}}) = \omega_{\hat{p}_i}\, X^i_{\hat{p}}, \quad \omega_{\hat{p}} = \omega_{\hat{p}_i}\, dx^i|_{\hat{p}}, \quad X_{\hat{p}} = X^i_{\hat{p}}\, \partial_i|_{\hat{p}},$$

where the index $\hat{p} \in \mathcal{M}$ indicates the basepoint.

From Eq. (4.19), we can define the differential 1-form field by

$$(\mathrm{d}f(X))(\hat{p}) = \mathrm{d}f|_{\hat{p}}(X_{\hat{p}}). \tag{5.6}$$

In the natural basis,

$$\mathrm{d}f = \frac{\partial f}{\partial x^i}\,\mathrm{d}x^i,$$

and in an arbitrary local basis $\{e_i\}_{i=1}^n$ in $\mathcal{T}_{\hat{p}}(\mathcal{M})$, where $\{\varepsilon^i\}_{i=1}^n$ is the dual basis, it follows that

$$(\mathrm{d}f(X))(\hat{p}) = X^i_{\hat{p}}\, e_i|_{\hat{p}}(f) = e_i|_{\hat{p}}(f)\, \varepsilon^i|_{\hat{p}}(X_{\hat{p}}),$$

and finally

$$\mathrm{d}f = e_i(f)\, \varepsilon^i. \tag{5.7}$$

5.4 Action of a Diffeomorphism

5.4.1 Image of a Vector Field Under a Diffeomorphism

Assume $\mathcal{M}$ and $\mathcal{M}'$ are two manifolds. Let

$$f : \mathcal{M} \to \mathcal{M}'$$

be a differentiable mapping between $\mathcal{M}$ and $\mathcal{M}'$, such that $\hat{p} \in \mathcal{M}$ is mapped to $p' = f(\hat{p}) \in \mathcal{M}'$.

The map f induces a linear (Jacobian) mapping[4] denoted $f_\star$, between the tangent space $\mathcal{T}_{\hat{p}}(\mathcal{M})$ at $\hat{p}$ and the tangent space $\mathcal{T}_{p'}(\mathcal{M}')$ at $p' = f(\hat{p})$:

$$f_\star : X_{\hat{p}} \mapsto f_\star(X_{\hat{p}}) = X'_{f(\hat{p})},$$

defined in the following way.

Let g be a differentiable function in the neighborhood of $f(\hat{p})$. Then we obtain

$$[f_\star(X)(g)](f(\hat{p})) \;=\; [X(g \circ f)](\hat{p}) \;=\; X_{\hat{p}}(g \circ f). \tag{5.8}$$

Theorem 5.2 *Let $f : \mathcal{M} \to \mathcal{M}'$ be a C^∞ diffeomorphism between two n-dimensional differentiable manifolds. Then $f_\star$ is an isomorphism of the Lie algebra:*

$$\mathfrak{T}(\mathcal{M}) \to \mathfrak{T}(\mathcal{M}),$$

given by

$$f_\star([X, Y]) = [f_\star(X), f_\star(Y)]. \tag{5.9}$$

Proof Exercise! □

Definition 5.15 (Invariant vector field) A vector field X on $\mathcal{M}$ is said to be invariant under the diffeomorphism $f : \mathcal{M} \to \mathcal{M}$ if

$$f_\star(X_{\hat{p}}) = X_{f(\hat{p})}, \quad \forall\, \hat{p} \in \mathcal{M}. \tag{5.10}$$

5.4.2 Image of a Cotangent Vector Field Under a Diffeomorphism

We consider a differentiable map $f : \mathcal{M} \to \mathcal{M}'$. Let $\hat{p} \in \mathcal{M}$ and define $p' = f(\hat{p}) \in \mathcal{M}'$.

The differential (push-forward) of f at $\hat{p}$ is a linear map:

$$f_* : \mathcal{T}_{\hat{p}}(\mathcal{M}) \to \mathcal{T}_{p'}(\mathcal{M}').$$

---

[4] The Jacobian mapping is sometimes denoted f'.

If $X_{\hat{p}} \in \mathcal{T}_{\hat{p}}(\mathcal{M})$, then its image is denoted

$$X'_{p'} = f_*(X_{\hat{p}}) \in \mathcal{T}_{p'}(\mathcal{M}').$$

The pull-back f^* maps a covariant vector (covector) at $p' \in \mathcal{M}'$ to one at $\hat{p} \in \mathcal{M}$:

$$f^* : \mathcal{T}^*_{p'}(\mathcal{M}') \longrightarrow \mathcal{T}^*_{\hat{p}}(\mathcal{M}).$$

Let $\omega_{\hat{p}} \in \mathcal{T}^*_{\hat{p}}(\mathcal{M})$ and $\omega_{p'} \in \mathcal{T}^*_{p'}(\mathcal{M}')$ be such that

$$\omega_{\hat{p}}(X_{\hat{p}}) \;=\; \omega_{p'}(X_{p'}),$$

where $X_{p'} = f_*(X_{\hat{p}})$.

The pull-back (reciprocal image)

$$f^* : \mathcal{T}^*_{p'}(\mathcal{M}') \;\longrightarrow\; \mathcal{T}^*_{\hat{p}}(\mathcal{M})$$

of a covariant vector $\omega_{p'}$ under the differentiable mapping f is defined by the equality

$$\left(f^*(\omega')\right)_{\hat{p}}(X_{\hat{p}}) \;=\; \omega_{p'}\left(f_*(X_{\hat{p}})\right).$$

Remark 5.1 The expression of the reciprocal image of a 1-form does not involve f^{-1}, whereas it is the case for a vector field.

5.4.3 Tensor Field

We generalize the discussion in the previous chapter about tensors. This notion also can evolve using the concept of fiber bundles.

In particular, a fiber bundle where the base space is the manifold $\mathcal{M}$ and the fiber is identified with $\otimes^p \mathcal{T}^\star_{\hat{p}}(\mathcal{M}) \otimes^q \mathcal{T}_{\hat{p}}(\mathcal{M})$ at all $\hat{p} \in \mathcal{M}$ is called a (p, q)-*tensor bundle*.

Notice that a tangent bundle is a $(0, 1)$-tensor bundle and that the cotangent bundle is a $(1, 0)$-tensor bundle.

Definition 5.16 A (p, q)-tensor field on a C^k-manifold is a C^r cross-section, where $r \leq k-1$ of the (p, q)-tensor bundle.

Operations defined in Sect. 4.5 for tensors at a point, are carried over fiber-wise allowing to define similar operations on tensor fields.

Under these operations, the set of all (p, q)-tensor fields of class C^r is a *module on the ring* $C^r(\mathcal{M})$.

Connections, Parallel Transport, and Sheafs 6

In differential geometry, connections and parallel transport arise from a fundamental necessity: the ability to compare vectors at different points on a curved space, where a naive translation is no longer well-defined.

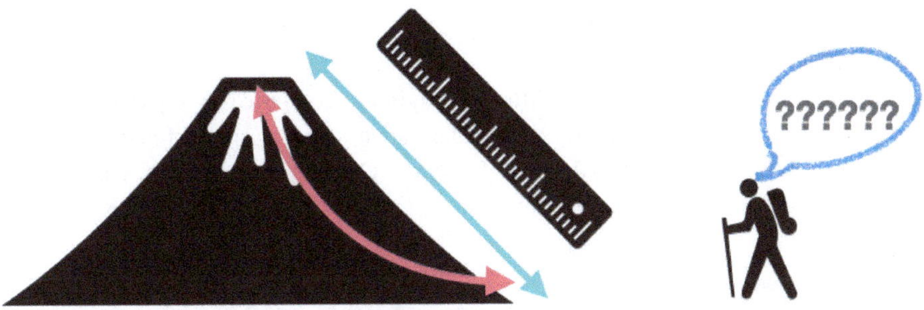

In Euclidean space, one can simply compare vectors via a parallel transport–shifting, without change in magnitude or direction.

However, on a curved manifold, the very act of moving a vector must be prescribed by an additional structure, which is provided by a connection.

A connection defines a rule for differentiation that respects the manifold's geometry, replacing the notion of partial derivative in a given direction with a well-defined covariant derivative. This operator tells us how a vector field changes as we move along the manifold, encoding information about curvature and torsion.

But the true power of connections emerges in the concept of parallel transport. Given a connection, we can transport a vector along a curve, according to the manifold's structure. This operation is at the heart of many geometric investigations.

In Riemannian geometry, parallel transport reveals the curvature of space, since moving a vector along different paths can lead to different final results. In fact, given a loop

© The Author(s), under exclusive license to Springer Nature Switzerland AG 2026

N. C. Combe et al., *Exploring Information Geometry*, Frontiers in Mathematics,
https://doi.org/10.1007/978-3-032-23054-6_6

$\gamma : [a, b] \to \mathcal{M}$ where $[a, b] \subset \mathbb{R}$ on a smooth manifold $\mathcal{M}$, where $\gamma(a) = \gamma(b)$ we can transport a vector $\mathbf{v}$ along that curve, starting at the initial point $\gamma(a)$.

The question that emerges is whether there has been any modification implied on the vector $\mathbf{v}$, during this transport, namely does the vector at $\gamma(b)$ coincide with the vector $\mathbf{v}$ at $\gamma(a)$ or not. If those vectors coincide, then the manifold is flat on the set where the loop was defined.

In gauge theory, parallel transport describes how fields interact, with connections playing the role of gauge potentials in Yang–Mills theory.

In general relativity, the Levi–Civita connection governs the motion of free-falling observers in curved spacetime.

Thus, a connection is not merely a technical tool but an essential ingredient of geometry, allowing us to define differentiation, curvature, and transport in a way that transcends local coordinates. It is the bridge between the infinitesimal and the global, between algebra and topology, between abstract geometry and the laws of nature.

> Warning: In what follows, we use the Einstein summation convention.

6.1 Linear Connection

In Euclidean space, two vectors of different origin are compared by a parallel translation of the vectors to the one same origin. The derivative of a vector $\mathbf{v}$ defined along a curve is a vector of components $(\partial v^i/\partial x^j)\mathrm{d}x^j/\mathrm{d}t$ in the Cartesian coordinates. On an arbitrary differentiable manifold, the components

$$\partial v^i/\partial x^j$$

do not behave as the components of a tensor under a change of coordinates. To solve this difficulty, we introduce the notion of covariant derivative which is of tensorial type.

Definition 6.1 (Linear connection) A linear connection on a differentiable manifold $\mathcal{M}$ is a mapping $\nabla : X \mapsto \nabla X$ from the vector fields on $\mathcal{M}$ into the differentiable tensor field of type $(1, 1)$ on $\mathcal{M}$ such that

$$\begin{aligned} \nabla(X+Y) &= \nabla(X) + \nabla(Y), \\ \nabla(fX) &= \mathrm{d}f \otimes X + f\nabla(X), \end{aligned} \tag{6.1}$$

where f is a differentiable function[1] on $\mathcal{M}$.

[1] To give a more precise definition, we have to speak of germ of vector (or tensor) field and germ of function on $\mathcal{M}$.

The tensor ∇X is called the *covariant derivative* of X.

For a n-dimensional differentiable manifold $\mathcal{M}$, equipped with moving a frames (Definition 5.12) $\{e_i\}_{i=1}^n$, $e_i \in \mathcal{T}(\mathcal{M})$ and $\{\varepsilon^i\}_{i=1}^n$, $\varepsilon^i \in \mathcal{T}^\star(\mathcal{M})$ (its dual), we have that ∇e_i is a $(1, 1)$-tensor field which can be written as:

$$\nabla e_i(e_k, \varepsilon^j) = \gamma_{ki}^j = (\nabla e_i)(\varepsilon^j, e_k), \tag{6.2}$$

and the $(1, 1)$-tensor ∇e_i can be expressed as:

$$\nabla e_i = \gamma_{ki}^j e_j \otimes \varepsilon^k. \tag{6.3}$$

Remark 6.1 For any covector $\omega \in \mathcal{T}_p^*\mathcal{M}$ and vector $Y \in \mathcal{T}_p\mathcal{M}$, the evaluation of the $(1, 1)$-tensors $\varepsilon^k \otimes e_i$ and $e_i \otimes \varepsilon^k$ is identical:

$$\boxed{(\varepsilon^k \otimes e_i)(\omega, Y) = (e_i \otimes \varepsilon^k)(\omega, Y)}.$$

Thus, $\nabla X = (\nabla_k X^i)\varepsilon^k \otimes e_i$ is identical to $\nabla X = (\nabla_k X^i)e_i \otimes \varepsilon^k$.

Definition 6.2 (Connection coefficients) The coefficients γ_{ki}^j are called the connection coefficients in the basis $\{e_i\}_{i=1}^n$.

It follows from Definition 6.1 and Eq. (4.18) that

$$\begin{aligned}
\nabla X &= \nabla(X^i e_i) = \mathrm{d}X^i \otimes e_i + X^i \nabla(e_i) \\
&= \left(\mathrm{d}X^i + \gamma_{kj}^i X^j \varepsilon^k\right) \otimes e_i \\
&= \left(e_k(X^i) + \gamma_{kj}^i X^j\right)\varepsilon^k \otimes e_i \\
&= \nabla_k X^i \varepsilon^k \otimes e_i,
\end{aligned}$$

where $\nabla_k X^i = \left(\nabla X\right)_k^i$ denote the component of the tensor ∇X.

$$\nabla X = \nabla_k X^i \varepsilon^k \otimes e_i, \quad \text{with} \quad \nabla_k X^i = e_k(X^i) + \gamma_{kj}^i X^j. \tag{6.4}$$

Furthermore, one can rewrite the $(1, 1)$-tensor ∇_X in the following way:

$$\nabla X = \left(\mathrm{d}X^i + \omega_j^i X^j\right) \otimes e_j, \quad \omega_j^i = \gamma_{kj}^i \varepsilon^k \tag{6.5}$$

making the covariant part explicit.

Definition 6.3 (Connection forms) The quantities defined by

$$\omega_j^i = \gamma_{kj}^i \varepsilon^k \tag{6.6}$$

are called the connection 1-forms.

We have the relation

$$\omega^i_j(e_k) = \gamma^i_{kj} = \nabla e_j(e_k, \varepsilon^i). \tag{6.7}$$

As ∇X is a $(1, 1)$-tensor, it is invariant under the change of natural coordinate system. In particular we have basis transformation (where the matrix A is invertible but not necessarily assumed orthogonal):

$$e_i = A^p_i e'_p, \quad \varepsilon^k = A^k_m \varepsilon'^m, \tag{6.8}$$

and additionally

$$A^\ell_i B^j_\ell = \delta^j_i, \quad B^\ell_i A^j_\ell = \delta^j_i, \tag{6.9}$$

where $B = A^{-1}$ is the inverse matrix such that

$$e'_j = B^k_j e_k, \quad \varepsilon'^\ell = B^\ell_m \varepsilon^m. \tag{6.10}$$

Basis Transformation

Given the basis transformation presented in Eq. (i.e., 6.8 $e_i = A_i^{\;p}\, e'_p$) and the vector expansion

$$X = X^i e_i = X'^p e'_p,$$

we substitute the basis so as to obtain the following:

$$X^i \underbrace{A_i^{\;p} e'_p}_{=e_i} = X'^p e'_p,$$

which implies that $X^i A_i^{\;p} = X'^p$ and given that

$$X^i = (A^{-1})^i{}_p\, X'^p = B^i{}_p X'^p,$$

by formula 6.9.

Therefore,

$$X^j = B^j_n X'^n \quad \text{implies} \quad \mathrm{d}X^i = B^i_j \mathrm{d}X'^j + X'^j \mathrm{d}B^i_j. \tag{6.11}$$

It follows from (6.5) that

$$\nabla X = \left(\mathrm{d}X^i + \gamma^i_{kj} X^j \varepsilon^k\right) \otimes e_i = \left(\mathrm{d}X'^j + \gamma'^j_{k\ell} X'^\ell \varepsilon'^k\right) \otimes e'_j.$$

Moreover,

$$\begin{aligned}
\nabla X &= \left(dX^i + \gamma^i_{kj}X^j\varepsilon^k\right)\otimes e_i \\
&= \Big[\underbrace{B^i_j dX'^j + X'^j dB^i_j + \gamma^i_{kj}(B^j_\ell X'^\ell)}_{\text{Eq. 6.11}}\underbrace{(A^k_m \varepsilon'^m)}_{\text{Eq. 6.8}}\Big]\otimes \underbrace{(A^p_i e'_p)}_{\text{Eq. 6.8}} \\
&= A^p_i\left(B^i_j dX'^j + X'^j dB^i_j + \gamma^i_{kj}B^j_\ell A^k_m X'^\ell \varepsilon'^m\right)\otimes e'_p \\
&= \underbrace{\delta^p_j dX'^j}_{\text{Eq. 6.9}}\otimes e'_p + A^p_i\left(X'^j dB^i_j + \gamma^i_{kj}B^j_\ell A^k_m X'^\ell \varepsilon'^m\right)\otimes e'_p \\
&= dX'^p\otimes e'_p + \underbrace{\left[A^p_i e'_h(B^i_j) + A^p_i A^k_h \gamma^i_{k\ell}B^\ell_j\right]}_{=\gamma'^j_{h\ell}} X'^j\varepsilon'^h\otimes e'_p
\end{aligned}$$

and therefore, by Eq. (4.19) we obtain:

$$\gamma'^j_{h\ell} = A^j_i e'_h(B^i_\ell) + A^j_i A^k_h B^m_\ell \gamma^i_{km}. \tag{6.12}$$

The connection coefficients are not components of a tensor because of the first term on the right-hand side of Eq. (6.12). The connection coefficients are not components of a tensor because of the first term on the right-hand side of Eq. (6.12).

In the local coordinate system $\{x^1, \dots, x^n\}$, we have:

$$\nabla X = \left(\frac{\partial X^i}{\partial x^k} + \Gamma^i_{k\ell}X^\ell\right)\mathrm{d}x^k \otimes \frac{\partial}{\partial x^i}, \tag{6.13}$$

where $\Gamma^i_{k\ell}X^\ell$ are the connection coefficients (6.3) in the local (natural) coordinate system $\{x^1, \dots, x^n\}$. For the vector field $\partial_i = \frac{\partial}{\partial x^i}$, the components are all 0 except for $X^i = 1$, and we can express:

$$\nabla(\partial_i) = \Gamma^j_{ki}\mathrm{d}x^k \otimes \partial_j.$$

We draw the reader's attention to the notation we used above, $\nabla(\partial_i)$. This notation indicates that ∇ is applied to the vector field ∂_i. While such notation is usually omitted in the literature, we prefer to retain it here for reasons of clarity.

Definition 6.4 (Christoffel symbols) The Γ^j_{ki} are called the Christoffel symbols. They are defined by:

$$\Gamma^j_{ki} = \nabla(\partial_i)(\partial_k, \mathrm{d}x^j) = \omega^j_i(\partial_k), \tag{6.14}$$

where:

- $\nabla(\partial_i)$ is the covariant derivative of the basis vector ∂_i (see Sect. 6.1.1).

- $\nabla(\partial_i)(\partial_k, dx^j)$ extracts the (k, j)-component of the $(1, 1)$-tensor $\nabla(\partial_i)$.
- $\omega_i^j(\partial_k)$ evaluates the connection 1-form ω_i^j at ∂_k.

Now, the Christoffel symbols transform under a change of local coordinates system: $\{x^1, \ldots, x^n\} \to \{x'^1, \ldots, x'^n\}$ as

$$\Gamma'^j_{h\ell} = \frac{\partial x'^j}{\partial x^i} \frac{\partial^2 x^i}{\partial x'^h \partial x'^\ell} + \frac{\partial x'^j}{\partial x^a} \frac{\partial x^b}{\partial x'^h} \frac{\partial x^c}{\partial x'^\ell} \Gamma^a_{bc}, \tag{6.15}$$

where $\frac{\partial x^b}{\partial x'^h}$ and $\frac{\partial x^c}{\partial x'^\ell}$ are inverse Jacobian components.

For any quantity to be a tensor, its transformation law between coordinate systems must be homogeneous and multilinear in the Jacobian matrices. The presence of the second-derivative term $\frac{\partial^2 x^i}{\partial x'^h \partial x'^\ell}$ confirms that Christoffel symbols do not transform as tensors:

$$\Gamma'^j_{h\ell} = \underbrace{\frac{\partial x'^j}{\partial x^a} \frac{\partial x^b}{\partial x'^h} \frac{\partial x^c}{\partial x'^\ell} \Gamma^a_{bc}}_{\text{(tensorial part)}} + \underbrace{\frac{\partial x'^j}{\partial x^i} \frac{\partial^2 x^i}{\partial x'^h, \partial x'^\ell}}_{\text{(non-tensorial term)}}.$$

Hence, the Christoffel symbols do not define a tensor.

Christoffel symbols Γ'^i_{jk} encode how the basis vectors change under parallel transport (they are the connection coefficients). The appearance of the non-tensorial second-derivative term corresponds to:

- **Curvature-Induced Effects**: Local bending of the manifold influences how vectors are compared at nearby points.
- **Path Dependence of Parallel Transport**: Different curves between the same endpoints yield different transports.

Their non-tensorial character signifies that they describe how *coordinate systems* twist and deform relative to parallel transport, rather than representing intrinsic geometric objects.

Affine Transformations in Flat Space

An affine transformation in flat space is defined as:

$$x'^j = M^j_i x^i + b^j,$$

where M^j_i is a constant invertible matrix (linear part) and b^j is a constant vector (the translation).

The inverse transformation is

$$x^i = (M^{-1})^i_j (x'^j - b^j).$$

There first derivative is

$$\frac{\partial x^i}{\partial x'^h} = (M^{-1})^i{}_h = \text{constant},$$

and the second derivative vanishes:

$$\frac{\partial^2 x^i}{\partial x'^h \, \partial x'^\ell} = 0,$$

so that the non-tensorial term vanishes. In curved manifolds (spacetime with gravity), no global affine coordinates exist. Coordinate transformations are generally non-affine which means that x'^j is not linear in x^i and that the second derivative is non null.

Remark 6.2 This plays a central role in the notion of Frobenius manifolds (see Chap. 10 on this topic).

6.1.1 Covariant Derivative of a Vector Field in the Direction of Y

Let X and Y be two vector fields. By (6.13), they can be expressed as

$$\nabla X(Y) = Y^k \nabla_k X^i e_i = Y^k \nabla_k X,$$

which can be identified to the derivative (4.3) in the direction of Y. This leads us to the formulation of a definition of the covariant derivative.

Definition 6.5 (Covariant derivative in the direction Y**)** The covariant derivative $\nabla_Y X$ of X in the direction Y is defined by

$$\nabla_Y X = (\nabla(X))(Y) \text{ or } \nabla_Y X(\cdot) = (\nabla(X))(Y, \cdot). \tag{6.16}$$

The covariant derivative $\nabla_Y X$ is *linear* in Y

$$\nabla_{fY_1 + gY_2} X = (f \nabla_{Y_1} + g \nabla_{Y_2}) X,$$

where

$$Y_1, Y_2 \in \mathcal{T}_p(\mathcal{M}), \quad \text{and} \quad f, g : \mathcal{M} \to \mathbb{R}.$$

In the frame $\{e_i\}_{i=1}^n$, we obtain:

$$\nabla_Y X = Y^k \left(e_k(X^i) + \gamma^i_{kj} X^j \right) e_i = \left(Y(X^i) + \gamma^i_{kj} Y^k X^j \right) e_i,$$

and thus

$$\nabla_{e_i} e_k = \gamma^j_{ik} e_j, \tag{6.17}$$

where ∇_{e_i} denote the covariant derivative in the direction e_i.

6.1.2 The Covariant Derivative of a Tensor

The notion of a covariant derivative in the direction of $X \in \mathcal{T}_{\hat{p}}(\mathcal{M})$ can be extended—in a neighborhood of $\hat{p} \in \mathcal{M}$—to an arbitrary type of tensor field, under the following assumptions:

$$\nabla_X f = X(f), \quad f \in C^1(\mathcal{M}), \tag{6.18a}$$

$$\nabla_X(t + s) = \nabla_X t + \nabla_X s, \tag{6.18b}$$

$$\nabla_X(t \otimes s) = (\nabla_X t) \otimes s + t \otimes (\nabla_X s), \tag{6.18c}$$

$$\nabla_X \text{ commutes with the contracted multiplication.} \tag{6.18d}$$

Therefore, if t is a (p, q)-tensor, its covariant derivative ∇t is a $(p, q + 1)$-tensor:

$$(\nabla t)(\omega_1, \ldots, \omega_p, v_1, \ldots, v_q, u) = (\nabla_u t)(\omega_1, \ldots, \omega_p, v_1, \ldots, v_q),$$

where

$$\omega_j \in \mathcal{T}^*_{\hat{p}}(\mathcal{M}), \quad u, v_k \in \mathcal{T}_{\hat{p}}(\mathcal{M}).$$

(a) Covariant Derivative of a 1-Form

We will study the above definition on the example of a 1-form. By the fourth condition above, we have that

$$\nabla_Y[\alpha(X)] = (\nabla_Y \alpha)(X) + \alpha(\nabla_Y X),$$

which implies that

$$(\nabla_Y \alpha)(X) = \nabla_Y[\alpha(X)] - \alpha(\nabla_Y X).$$

If we substitute X by e_i, this gives

$$(\nabla_Y \alpha)_i = Y(\alpha_i) - \alpha(\gamma^j_{ki} Y^k e_j) = Y(\alpha_i) - \gamma^j_{ki} Y^k \alpha_j,$$

and thus

$$\nabla_Y \alpha = Y^k \left[e_k(\alpha_i) - \gamma^j_{ki} \alpha_j \right] \varepsilon^i.$$

For the dual basis ε^i (where $\alpha = \varepsilon^i$), we have $\alpha_i = \delta^i_i = 1$ and $\alpha_j = \delta^i_j$, so:

$$\nabla_Y \varepsilon^i = Y^k \left[e_k(\delta^i_j) - \gamma^j_{ki}\delta^i_j \right] \varepsilon^i = -Y^k \gamma^i_{ki} \varepsilon^i .$$

The last expression is:

$$\nabla_Y \varepsilon^i = -Y^k \gamma^i_{kj} \varepsilon^j \quad \text{or, in other words} \quad \nabla_{e_k} \varepsilon^i = -\gamma^i_{kj} \varepsilon^j . \tag{6.19}$$

Proposition 6.1 (Covariant derivative of 1-form) *Let $\alpha \in \mathcal{T}^\star_p(\mathcal{M})$. Then*

$$(\nabla_Y \alpha)(X) = \nabla_Y(\alpha X) - \alpha(\nabla_Y X). \tag{6.20}$$

Additionally, if $\{e_i\}^n_{i=1}$ is a moving frame and $\{\varepsilon^i\}^n_{i=1}$, then

$$\nabla \alpha = (d\alpha_i + \alpha_j \omega^j_i) \otimes \varepsilon^i \tag{6.21}$$

Proof The proof is left as an exercise. □

(b) Covariant Derivative of a Tensor

We now consider a (0, 2)-tensor and explore the covariant derivative of a (0, 2)-tensor. Let $g = g_{\mu\nu}\varepsilon^\mu \otimes \varepsilon^\nu$ be a (0, 2)-tensor.

The covariant derivative of g is obtained from the properties (3) and (4) outlined in the list of conditions above.

Therefore,

$$\begin{aligned} \nabla_X g &= X(g_{\mu\nu})\epsilon^\mu \otimes \epsilon^\nu + g_{\mu\nu}\nabla_X \varepsilon^\mu \otimes \epsilon^\nu + g_{\mu\nu}\varepsilon^\mu \otimes \nabla_X \varepsilon^\nu \\ &= X^\alpha [e_\alpha(g_{\mu\nu}) - \gamma^\beta_{\alpha\mu} g_{\beta\nu} - \gamma^\beta_{\alpha\nu} g_{\mu\beta}]\varepsilon^\mu \otimes \varepsilon^\nu , \end{aligned}$$

or

$$\nabla_\alpha g_{\mu\nu} = e_\alpha(g_{\mu\nu}) - \gamma^\beta_{\alpha\mu} g_{\beta\nu} - \gamma^\beta_{\alpha\nu} g_{\mu\beta}. \tag{6.22}$$

Exercise 6.1 The generalization is rather straight forward and left as an exercise.

Exercise 6.2 As an example, one can work with a (2, 1)-tensor t^i_{kl} and show that:

$$\nabla_j t^i_{k\ell} = e_j(t^i_{k\ell}) + \gamma^i_{jm} t^m_{k\ell} - \gamma^m_{jk} t^i_{m\ell} - \gamma^m_{j\ell} t^i_{km}.$$

We continue with the next remark.

Remark 6.3 The covariant derivative in the direction e_i of the tensor product of s and t of two tensors satisfy, by (6.18c):

$$\nabla_{e_i}(s \otimes t) = (\nabla_{e_i} s) \otimes t + s \otimes (\nabla_{e_i} t).$$

However, it is important to keep in mind that the sum of two tensor products is defined only if the corresponding factors have the same rank and therefore:

$$\nabla(s \otimes t) \neq (\nabla s) \otimes t + s \otimes (\nabla t).$$

6.2 Parallel Transport—Geodesics

In Euclidean space, vectors based at different points can be compared by translating them parallelly to a common origin.

Consequently, if a vector $\mathbf{v}$ is transported parallelly along a curve γ, the derivative $\mathrm{d}v/\mathrm{d}t = 0$ vanishes.

On a manifold $\mathcal{M}$, to compare vectors at two different points $M \in \mathcal{M}$ and $M' \in \mathcal{M}$, it is necessary to be able to assign a uniquely defined frame at M' based on a frame at $\mathcal{M}$.

This is precisely where the notion of connection plays a key role: By enabling the existence of a covariant derivative, a connection provides a convenient notion of parallel transport.

6.2.1 Parallel Vector Along a Curve

Definition 6.6 (Parallel transport) A vector X is said to be parallel along the curve $\gamma : t \to \gamma(t)$ if

$$\nabla_u X = 0, \quad u = \gamma' \frac{\mathrm{d}}{\mathrm{d}t}. \tag{6.23}$$

Remark 6.4 The vector u is defined only at points lying on the curve γ. Somehow, it can also be extended to a vector field on a neighborhood of the curve γ.

If (φ, U) is a local chart, the component of a point $\gamma(t)$ on the curve γ is

$$x^i(t) = \varphi^i \circ \gamma(t)$$

and the components of u are given by $u^i = \mathrm{d}x^i/\mathrm{d}t$

$$u^i = \varphi^i \circ u = u(\varphi^i).$$

6.2.2 Geodesics

We shortly digress on the notion of geodesics, as they intimately related to the topic discussed in this section. The concept of geodesics generalizes the notion of straight line in Euclidean space.

Definition 6.7 (Affine geodesic) A curve $\gamma : I \to \mathcal{M}$ is a *pregeodesic* if it satisfies

$$\nabla_u u = \lambda(t)u, \quad u = \frac{d\gamma}{dt}, \tag{6.24}$$

for some $\lambda(t)$. It becomes an *affine geodesic* (with affine parameter) when reparametrized to satisfy $\nabla_u u = 0$.

The formula in Eq. 6.24 captures the essence of geodesic paths (straightness) independent of parametrization. The additional condition that $\nabla_u u = 0$ selects a canonical parametrization.

Consider a smooth curve $\gamma(t)$ on a manifold $\mathcal{M}$ with tangent vector field $u = \dot{\gamma}(t)$. The equation

$$\nabla_u u = \lambda(t)u$$

expresses that the acceleration $\nabla_u u$ is parallel to the velocity u. This has the following geometric interpretation:

- The curve $\gamma(t)$ traces a "straight" path in the geometry of $\mathcal{M}$, although its parametrization may be non-uniform.
- The curve is a *pregeodesic*: It coincides with a geodesic as a set, but its speed may vary with t.

△ **Physical Necessity**
In physics, geodesics describe the trajectories of free particles (e.g., planetary motion in general relativity). The condition

$$\nabla_u u = \lambda(t)u$$

permits arbitrary parametrizations (for example, coordinate time t instead of proper time τ), while ensuring that the underlying path remains physically meaningful. Thus, whether a planet's orbit is parametrized by time or by arc length, it represents the same physical trajectory.

△ **Mathematical Necessity**
The freedom to reparametrize is essential:

- Solutions to $\nabla_u u = \lambda(t)u$ are invariant under changes of parameter $t \mapsto \tau(t)$, whereas the strict geodesic equation $\nabla_u u = 0$ is not.

- Pregeodesics ($\nabla_u u = \lambda u$) define the same geometric path as true geodesics ($\nabla_u u = 0$), after reparametrization.
- For every pregeodesic there exists an affine parameter $s(t)$ such that the reparametrized curve $\tilde{\gamma}(s) = \gamma(t(s))$ satisfies

$$\nabla_{\tilde{u}} \tilde{u} = 0, \qquad \text{where } \tilde{u} = \frac{\mathrm{d}\tilde{\gamma}}{\mathrm{d}s}.$$

Consequences of the Condition
If $\nabla_u u = \lambda u$, then:

- The curve is *autoparallel*: Its velocity vector u is parallel-transported along the curve up to scaling.
- The image of the curve in $\mathcal{M}$ coincides with that of a geodesic.
- An affine parameter s can be constructed by solving

$$\frac{\mathrm{d}^2 s}{\mathrm{d}t^2} + \lambda \frac{\mathrm{d}s}{\mathrm{d}t} = 0,$$

such that, in the new parametrization, the geodesic equation takes the form

$$\nabla_{\tilde{u}} \tilde{u} = 0.$$

There are some easy computations that can be done given $\gamma : I \to \mathcal{M}, \quad I = [a, b] \subset \mathbb{R}$, the length of the smooth curve is given by

$$L(\gamma) := \int_a^b |\dot{\gamma}(t)| \mathrm{d}t.$$

In particular, this notions starts to be interesting as soon as we want to understand better the intrinsic distance between two points on a manifold. Assume $\mathcal{M} \subset \mathbb{R}^n$ is an $\mathcal{M}$-dimensional smooth manifold. Take a pair of distinct points $p, q \in \mathcal{M}$. Their Euclidean distance is given by $|p - q|$ in the ambient Euclidean space. However, this type of distance does not tell us much, from the viewpoint of the manifold $\mathcal{M}$. We therefore introduce the notion of an intrinsic distance in $\mathcal{M}$.

Definition 6.8 The intrinsic distance on $\mathcal{M}$ between $p, q \in \mathcal{M}$ is a real number $d(p, q) \geq 0$ defined by

$$d(p, q) := \inf_{\gamma \in \Omega_{p,q}} L(\gamma),$$

where $\Omega_{p,q}$ is the space of smooth paths of the unit interval joining p to q.

Exercise 6.3 Prove that $L(\gamma) \geq |p - q|$.

Let us now write the equation of geodesics in local chart (φ, U) the component of a point $\gamma(t)$ on the curve γ are $x^i(t) = \varphi \circ \gamma(t)$ and the component of u are $u^i = \mathrm{d}x^i/\mathrm{d}t$.

6.3 Curvature

6.3.1 Vector Bundle—Version

Let $(\mathcal{M}, g)$ be a Riemannian manifold. The Riemann curvature tensor is the multilinear map

$$R : \Gamma(\mathcal{T}(\mathcal{M})) \times \Gamma(\mathcal{T}(\mathcal{M})) \times \Gamma(\mathcal{T}(\mathcal{M})) \to \Gamma(\mathcal{T}(\mathcal{M}))$$

defined by:

$$R(X, Y)Z = \nabla_X \nabla_Y Z - \nabla_Y \nabla_X Z - \nabla_{[X,Y]} Z,$$

where:

- $X, Y, Z \in \Gamma(\mathcal{T}(\mathcal{M}))$ are smooth vector fields.
- ∇ is the Levi–Civita connection.
- $[X, Y] = XY - YX$ is the Lie bracket.

This definition treats R as a $(1, 3)$ tensor where there is an input of three vectors (say X, Y, Z) and one output of one vector $(R(X, Y)Z)$.

Example 6.1 In $\mathbb{R}^n$ with the Euclidean metric:

- $\nabla_X Y$ is the directional derivative $D_X Y$.
- $R(X, Y)Z = D_X D_Y Z - D_Y D_X Z - D_{[X,Y]} Z \equiv 0$ (by equality of mixed partials).
- Geodesics are straight lines: Parallel transport preserves vectors.

Table 6.1 Comparison of fiber bundles and sheaves

1. Fiber bundles: Geometry with local triviality	2. Sheaves: Local-to-global algebraization
Origin: Arising from differential geometry/topology (e.g., tangent bundles, vector bundles)	*Motivation*: Jean Leray (1940s) introduced them to study PDEs and topology; further developed by Cartan, Serre, and Grothendieck
Key idea: A space E (total space) projecting onto a base B where each fiber $F_b = \pi^{-1}(b)$ is homeomorphic to a fixed fiber F, with **local triviality**: $E\|_U \cong U \times F$ for open sets $U \subset B$	*Key idea*: Attach algebraic data (rings, modules, groups) to open sets with restriction maps satisfying **gluing conditions**
Example: Tangent bundle $\mathcal{T}(\mathcal{M})$ of a manifold $\mathcal{M}$	*Example*: $\mathcal{O}_X$ (sheaf of holomorphic functions on a complex manifold)
Limitations: • Requires smooth/topological structure • Global properties depend on transition functions • Struggles with singular spaces/algebraic contexts	*Advantages*: • Handles non-free modules (torsion) and singularities • Directly encodes local data (germs of functions) • Sheaf cohomology ($H^i(X, \mathcal{F})$) generalizes classical cohomologies

6.3.2 Sheaf Version

NB The purpose of this section is to convey the flavor of the various techniques and constructions related to bundles, with an emphasis on their scope and utility. Our intention is not to present a complete or technically exhaustive treatment, but rather to provide an overview that keeps the reader informed of the range of available methods without interrupting the flow of the exposition. A comprehensive and detailed account of these constructions will be presented in our forthcoming textbook.

In the scope of using a more modern language, we can consider the notion of Riemannian curvature in a more categorical framework, which relies on the notion of sheaf. To give a rough idea, a sheaf assigns some local data to open sets and patches these local data together into a global object. For clarity, we present Table 6.1 highlighting the principal distinctions between the two notions.

The notion of presheafs can be described in purely set-theoretic terms as follows.

A *presheaf* $\mathcal{F}$ on a topological space X serves as a device for organizing local data over open subsets of X. More precisely:

- **Data Assignment**. To every open set $U \subseteq X$ one assigns a mathematical object $\mathcal{F}(U)$ (for instance, a set, group, or ring), which represents the admissible data defined on U.

- **Restriction**: For each inclusion $V \subseteq U$, a restriction map is defined

$$\mathrm{res}_{U,V} : \mathcal{F}(U) \to \mathcal{F}(V),$$

satisfying:

1. $\mathrm{res}_{U,U} = \mathrm{id}_{\mathcal{F}(U)}$ (identity),
2. if $W \subseteq V \subseteq U$, then

$$\mathrm{res}_{V,W} \circ \mathrm{res}_{U,V} = \mathrm{res}_{U,W}$$

(transitivity).

Example. For each open $U \subseteq X$, let $\mathcal{F}(U)$ be the ring of continuous functions on U, with restriction maps given by $f \mapsto f|_V$, where $V \subseteq U$.

Though optional, a categorical version of presheafs provides a unifying framework with deep advantages. We consider the notion of categories later in the book (see Sect. 8.1.3), used in the context of information geometry. In the footnote[2] we recall the basic definition, for convenience. Intuitively, one may think of a category as a network whose nodes are its objects and whose links are its morphisms. A functor between two categories $F : \mathcal{C} \to \mathcal{D}$ from the category $\mathcal{C}$ to $\mathcal{D}$ transports this entire network from $\mathcal{C}$ into $\mathcal{D}$ in a manner that preserves its structure: to each object X of $\mathcal{C}$ it assigns an object $F(X)$ of $\mathcal{D}$, and to each morphism $f : X \to Y$ in $\mathcal{C}$ it assigns a morphism $F(f) : F(X) \to F(Y)$ in $\mathcal{D}$. These assignments respect both composition and identities, in the sense that

$$F(g \circ f) = F(g) \circ F(f), \qquad F(\mathrm{id}_X) = \mathrm{id}_{F(X)}.$$

---

[2] A *category* is a mathematical structure encoding objects and the arrows between them. Formally, it consists of:

- **Objects**: Denoted by $A, B, C, \ldots$. Typical examples include sets, groups, topological spaces, or more abstract entities.
- **Morphisms (Arrows)**: For objects A and B, a morphism $f : A \to B$ is an arrow from A to B. The collection of all such morphisms is denoted by $\mathrm{Hom}(A, B)$.
- **Composition**: Given morphisms $f : A \to B$ and $g : B \to C$, there exists a composite morphism

$$g \circ f : A \to C,$$

satisfying the associativity law

$$(h \circ g) \circ f = h \circ (g \circ f).$$

- **Identity Morphisms**: Each object A carries a distinguished arrow $\mathrm{id}_A : A \to A$ such that, for any morphism $f : A \to B$,

$$f \circ \mathrm{id}_A = f \quad \text{and} \quad \mathrm{id}_B \circ f = f.$$

Warning! Categories exhibit several guiding principles that are important to keep in mind.

- **Structure over Elements**. The emphasis lies on the relationships (morphisms) between objects, rather than the internal composition of the objects themselves.
- **Universality**. Categories capture patterns that are invariant across mathematics, focusing on how structures interact rather than on what they contain.
- **Arrow-Centric Perspective**. Objects are characterized by the web of morphisms to and from them, making arrows the fundamental constituents of the theory.

The categorical version of presheafs goes as follows.

Definition 6.9 Let X be a topological space. A presheaf $\mathcal{F}$ on X is a contravariant functor[3] from the category of open sets of X (denoted $\mathrm{Open}(X)$), to a category $\mathscr{C}$, i.e.,

$$\mathcal{F} : \mathrm{Open}(X)^{\mathrm{op}} \to \mathscr{C}.$$

This definition means that to each open set $U \subset X$, we assign an object $\mathcal{F}(U)$ in the category $\mathscr{C}$. The category $\mathscr{C}$ can be the category of sets, abelian groups, rings, etc.

We now turn to an explanation of why the categorical formulation, despite being less immediate to intuition, ultimately reveals itself as the more powerful and unifying framework (Table 6.2).

A sheaf is a presheaf with an extra gluing condition. Namely, for any open covering $U = \bigcup_\alpha U_\alpha$ and a family of local sections that agree on overlaps (i.e., *compatible local sections*), there exists a unique global section that restricts to them. In particular, $\mathcal{F}$ is a sheaf if, for any open cover $U = \bigcup_\alpha U_\alpha$, the sequence is exact:

[3] Intuitively, a contravariant functor may be viewed as a rule that associates to each object of a category $\mathcal{C}$ an object of a category $\mathcal{D}$, while reversing the direction of all morphisms. Formally, a contravariant functor $F : \mathcal{C} \to \mathcal{D}$ consists of:

- an assignment $X \mapsto F(X)$ for every object X of $\mathcal{C}$;
- an assignment $f : X \to Y \mapsto F(f) : F(Y) \to F(X)$ for every morphism f in $\mathcal{C}$,

such that the following functoriality conditions hold:

$$F(g \circ f) = F(f) \circ F(g), \qquad F(\mathrm{id}_X) = \mathrm{id}_{F(X)}.$$

Table 6.2 Comparison between set-theoretic and categorical presheaves

Set-theoretic presheaves	Categorical presheaves
Bound to topology	Defined on any category $\mathcal{C}$
Ad hoc morphisms	Morphisms are natural transformations
Manual sheaf checks	Sheaf condition expressed as a limit
Limited to classical geometry	Powers schemes, topoi, and derived geometry

$$\mathcal{F}(U) \to \prod_{\alpha} \mathcal{F}(U_\alpha) \rightrightarrows \prod_{\alpha,\beta} \mathcal{F}(U_\alpha \cap U_\beta).$$

Let us breakdown this construction:

- The first map sends a global section $s \in \mathcal{F}(U)$ to its restriction $s|_{U_\alpha}$.
- The second pair of maps send (s_α) to their restrictions on overlaps $s_\alpha|_{U_\alpha \cap U_\beta}$ and $s_\beta|_{U_\alpha \cap U_\beta}$.

6.3.3 Curvature Version 2 (Algebraic)

The transition from classical differential geometry to modern sheaf-theoretic formalism involves these key reconceptualizations:

- The tangent bundle sections are viewed as a *sheaf* $\mathcal{E}$ of $\mathcal{O}_\mathcal{M}$-modules.
- The covariant derivative is a *morphism in the category* of sheaves rather than a pointwise operator.
- Curvature manifests as an *endomorphism-valued differential form.*
- The bracket operation is encoded in the *Lie algebroid structure* of $\mathcal{E}$.[4]

Let $(\mathcal{M}, g)$ be a Riemannian manifold with $\mathcal{O}_\mathcal{M}$ the sheaf of smooth functions and $\mathcal{E} = \mathcal{T}_\mathcal{M}$ the tangent sheaf. The connection is a morphism of $\mathcal{O}_\mathcal{M}$-modules:

$$\nabla : \mathcal{E} \to \Omega^1_\mathcal{M} \otimes \mathcal{E},$$

where $\Omega^1_\mathcal{M} = \mathrm{Der}(\mathcal{O}_\mathcal{M}, \mathcal{O}_\mathcal{M})$ is the sheaf of Kahler differentials. The *Riemann curvature tensor* is then defined as the $\mathcal{O}_\mathcal{M}$-linear morphism:

$$R : \mathcal{E} \otimes \mathcal{E} \to \mathrm{End}\,(\mathcal{E}) \otimes \Omega^2_\mathcal{M}$$

[4] A Lie algebroid is, roughly speaking, a vector bundle with a Lie algebra structure that acts on a manifold. Formally, it consists of a vector bundle $E \to \mathcal{M}$ whose space of sections $\Gamma(E)$ carries a Lie bracket, together with a bundle map (the *anchor*) $\rho : E \to \mathcal{T}(\mathcal{M})$ satisfying a Leibniz rule.

given explicitly by:

$$R(X, Y) = \nabla_X \nabla_Y - \nabla_Y \nabla_X - \nabla_{[X,Y]}$$

where $[X, Y]$ is the Lie bracket induced by the Lie algebroid structure of $\mathcal{E}$. This corresponds to a global section:

$$R \in \Gamma\left(\mathcal{M}, \operatorname{End}(\mathcal{E}) \otimes \bigwedge^2 \Omega^1_{\mathcal{M}}\right).$$

The curvature measures the *obstruction to flatness* of the connection through the commutator relation:

$$[\nabla_X, \nabla_Y] - \nabla_{[X,Y]} = R(X, Y)$$

and is $\mathcal{O}_{\mathcal{M}}$-multilinear in all arguments, confirming its tensorial nature. This formulation reveals curvature as a *functorial obstruction* to the integrability of the connection in the category of $\mathcal{O}_{\mathcal{M}}$-modules.

Since the right-hand side only depends on the values of X, Y, Z at a given point, R is a *tensorial* object, i.e., a section of the sheaf: $R \in \Gamma(\mathcal{M}, \operatorname{End}(\mathcal{E}) \otimes \Omega^2_{\mathcal{M}})$.

This tensor encodes the *non-commutativity* of the covariant derivative and measures the curvature of the category of $\mathcal{O}_{\mathcal{M}}$-modules endowed with ∇.

Exercise 6.4 Show that we can rewrite it as:

$$R(X, Y) = [\nabla_X, \nabla_Y] + \nabla_{[X,Y]}.$$

Exercise 6.5 Symmetries: Show that we have the following relations:

$$R(Y, X) = -R(X, Y) \tag{6.25}$$

$$R(X, Y)Z + R(Y, Z)X + R(Z, X)Y = 0 \tag{6.26}$$

$$\langle R(X, Y)Z, W\rangle = \langle R(Z, W)X, Y\rangle. \tag{6.27}$$

Note that we can rewrite the Riemannian curvature tensor differently using the universal property of tensor products and the Hom-tensor adjunction. The following discussion is primarily intended for readers who are new to the subject and wish to see detailed explanations; advanced readers may safely skip it, as the result follows from standard arguments.

Let $\mathcal{E}$ be a locally free $\mathcal{O}_{\mathcal{M}}$-module of finite rank (e.g., the tangent bundle on a manifold). The key is to recognize a different formulations arising from adjoint functors in the symmetric monoidal category of $\mathcal{O}_{\mathcal{M}}$-modules.

Step 1: Hom-Tensor Adjunction

The adjunction between $\otimes$ and Hom states that for $\mathcal{O}_{\mathcal{M}}$-modules A, B, C,

$$\text{Hom}_{\mathcal{O}_{\mathcal{M}}}(A \otimes B, C) \cong \text{Hom}_{\mathcal{O}_{\mathcal{M}}}\big(A, \text{Hom}_{\mathcal{O}_{\mathcal{M}}}(B, C)\big).$$

Here, $\text{Hom}_{\mathcal{O}_{\mathcal{M}}}(B, C)$ is the sheaf of $\mathcal{O}_{\mathcal{M}}$-linear morphisms.

Step 2: Relation Between Two Curvature Definitions
Set

$$A = \mathcal{E} \otimes \mathcal{E}, \quad B = \mathcal{E}, \quad C = \mathcal{E} \otimes \Omega^2_{\mathcal{M}}.$$

The adjunction gives:

$$\text{Hom}_{\mathcal{O}_M}(\mathcal{E} \otimes \mathcal{E} \otimes \mathcal{E}, \mathcal{E} \otimes \Omega^2_{\mathcal{M}}) \cong \text{Hom}_{\mathcal{O}_{\mathcal{M}}}\big(\mathcal{E} \otimes \mathcal{E}, \text{Hom}_{\mathcal{O}_{\mathcal{M}}}(\mathcal{E}, \mathcal{E} \otimes \Omega^2_{\mathcal{M}})\big). \tag{6.28}$$

Step 3: Simplify the Right-Hand Side
For any $\mathcal{O}_{\mathcal{M}}$-module $\mathscr{F}$, the tensor-hom adjunction states:

$$\text{Hom}_{\mathcal{O}_{\mathcal{M}}}(\mathcal{E}, \mathscr{F}) \cong \mathcal{E}^\vee \otimes \mathscr{F}$$

where $\mathcal{E}^\vee = \text{Hom}_{\mathcal{O}_{\mathcal{M}}}(\mathcal{E}, \mathcal{O}_{\mathcal{M}})$ is the dual sheaf. This isomorphism holds precisely because $\mathcal{E}$ is locally free of finite rank. Applying this to $\mathscr{F} = \mathcal{E} \otimes \Omega^2_{\mathcal{M}}$:

$$\text{Hom}_{\mathcal{O}_{\mathcal{M}}}(\mathcal{E}, \mathcal{E} \otimes \Omega^2_{\mathcal{M}}) \cong \mathcal{E}^\vee \otimes (\mathcal{E} \otimes \Omega^2_{\mathcal{M}}).$$

Furthermore, tensor products of $\mathcal{O}_M$-modules being associative, we obtain that:

$$\mathcal{E}^\vee \otimes (\mathcal{E} \otimes \Omega^2_{\mathcal{M}}) \cong (\mathcal{E}^\vee \otimes \mathcal{E}) \otimes \Omega^2_{\mathcal{M}}.$$

Now, since $\mathcal{E}$ is locally free of finite rank, we have that the endomorphism sheaf End $(\mathcal{E}) \cong \mathcal{E}^\vee \otimes \mathcal{E}$. Therefore,

$$(\mathcal{E}^\vee \otimes \mathcal{E}) \otimes \Omega^2_{\mathcal{M}} \cong \text{End}(\mathcal{E}) \otimes \Omega^2_{\mathcal{M}}.$$

Step 4: Apply the Isomorphism
Substituting into (6.28) yields:

$$\text{Hom}_{\mathcal{O}_{\mathcal{M}}}(\mathcal{E} \otimes \mathcal{E} \otimes \mathcal{E}, \mathcal{E} \otimes \Omega^2_{\mathcal{M}}) \cong \text{Hom}_{\mathcal{O}_M}(\mathcal{E} \otimes \mathcal{E}, \text{End}(\mathcal{E}) \otimes \Omega^2_{\mathcal{M}}). \tag{6.29}$$

Step 5: Account for Skew-Symmetry
The Riemann curvature tensor is skew-symmetric in its first two arguments. Define

$$K_1 = \left\{\varphi \in \text{Hom}_{\mathcal{O}_{\mathcal{M}}}(\mathcal{E} \otimes \mathcal{E}, \text{End}(\mathcal{E}) \otimes \Omega^2_{\mathcal{M}}) \mid \varphi(X, Y) = -\varphi(Y, X)\right\},$$

$$K_2 = \left\{\psi \in \text{Hom}_{\mathcal{O}_{\mathcal{M}}}(\mathcal{E} \otimes \mathcal{E} \otimes \mathcal{E}, \mathcal{E} \otimes \Omega^2_{\mathcal{M}}) \mid \psi(X, Y, Z) = -\psi(Y, X, Z)\right\}.$$

The isomorphism (6.29) restricts to an isomorphism

$$K_2 \cong K_1,$$

because the adjunction preserves the skew-symmetry in the first two arguments. Explicitly:

Given $\tilde{R} \in K_2$, define $R \in K_1$ by

$$[R(X, Y)](Z) := \tilde{R}(X, Y, Z).$$

Conversely, given $R \in K_1$, define $\tilde{R} \in K_2$ by

$$\tilde{R}(X, Y, Z) := [R(X, Y)](Z).$$

Step 6: Commutative Diagram
The equivalence is summarized in the diagram:

$$\begin{array}{ccc} K_1 & \xrightarrow{\sim} & K_2 \\ R : \mathcal{E} \otimes \mathcal{E} \to \mathrm{End}(\mathcal{E}) \otimes \Omega^2_{\mathcal{M}} & \longleftrightarrow & \tilde{R} : \mathcal{E} \otimes \mathcal{E} \otimes \mathcal{E} \to \mathcal{E} \otimes \Omega^2_{\mathcal{M}} \\ [R(X, Y)](Z) & \mapsto & \tilde{R}(X, Y, Z) \end{array}$$

with skew-symmetry in the first two arguments X, Y.

Conclusion
The universal property proves that the two definitions of the Riemann curvature tensor are equivalent:

$$\mathrm{Hom}_{\mathcal{O}_{\mathcal{M}}}(\mathcal{E} \otimes \mathcal{E}, \mathrm{End}(\mathcal{E}) \otimes \Omega^2_{\mathcal{M}})_{\mathrm{skew}} \cong \mathrm{Hom}_{\mathcal{O}_{\mathcal{M}}}(\mathcal{E} \otimes \mathcal{E} \otimes \mathcal{E}, \mathcal{E} \otimes \Omega^2_{\mathcal{M}}).$$

This isomorphism is independent of coordinates, fulfilling the universal property of tensor products.

6.3.3.1 Why Is the Fact That $\mathcal{E}$ Is Locally Free of Finite Rank Critical?

The isomorphism $\mathrm{Hom}_{\mathcal{O}_{\mathcal{M}}}(\mathcal{E}, \mathcal{E} \otimes \Omega^2_{\mathcal{M}}) \cong \mathrm{End}\,(\mathcal{E}) \otimes \Omega^2_{\mathcal{M}}$ depends critically on $\mathcal{E}$ being **locally free of finite rank**. We illustrate this fact on an example.

Assume $\mathcal{M} = \mathbb{R}^2$ (as a smooth manifold), and take $\mathcal{E} = \mathcal{T}(\mathcal{M})$ (tangent bundle), which is locally free of rank 2. Fix an open set $U \subset \mathbb{R}^2$ with coordinates (x, y).

Local Trivialization

$\mathcal{E}|_U \cong (\mathcal{O}_{\mathcal{M}}|_U)^{\oplus 2}$ with basis:

$$\mathbf{e}_1 = \frac{\partial}{\partial x}, \quad \mathbf{e}_2 = \frac{\partial}{\partial y}.$$

Every section $X \in \mathcal{E}(U)$ can be written as $X = f\mathbf{e}_1 + g\mathbf{e}_2$ for $f, g \in \mathcal{O}_{\mathcal{M}}(U)$.

Dual Basis

$\mathcal{E}^\vee|_U \cong (\mathcal{O}_{\mathcal{M}}|_U)^{\oplus 2}$ with dual basis:

$$\mathbf{e}^1 = \mathrm{d}x, \quad \mathbf{e}^2 = \mathrm{d}y$$

satisfying $\langle \mathbf{e}_i, \mathbf{e}^j \rangle = \delta_i^j$. For example:

$$\left\langle \frac{\partial}{\partial x}, \mathrm{d}x \right\rangle = 1, \quad \left\langle \frac{\partial}{\partial y}, \mathrm{d}x \right\rangle = 0.$$

Matrix Representation

Consider a homomorphism $\phi \in \mathrm{Hom}_{\mathcal{O}_{\mathcal{M}}}(\mathcal{E}, \mathcal{E} \otimes \Omega^2_{\mathcal{M}})$. Locally on U:

$$\phi(\mathbf{e}_1) = \alpha_{11}\mathbf{e}_1 \otimes \omega + \alpha_{21}\mathbf{e}_2 \otimes \omega$$

$$\phi(\mathbf{e}_2) = \alpha_{12}\mathbf{e}_1 \otimes \omega + \alpha_{22}\mathbf{e}_2 \otimes \omega,$$

where $\omega = \mathrm{d}x \wedge \mathrm{d}y \in \Omega^2_{\mathcal{M}}(U)$, and $\alpha_{ij} \in \mathcal{O}_{\mathcal{M}}(U)$. Thus ϕ is represented by the matrix:

$$\begin{pmatrix} \alpha_{11} & \alpha_{12} \\ \alpha_{21} & \alpha_{22} \end{pmatrix}.$$

We finally explain more globally why $\mathcal{E}$ needs to be locally free of finite rank.

- **Local Trivialization**: The condition $\mathcal{E}|_{U_i} \cong (\mathcal{O}_{\mathcal{M}}|_{U_i})^{\oplus n}$ allows us to work with **finite matrices**. For our tangent bundle example, $n = 2$ is finite, so:

$$\phi \leftrightarrow \begin{pmatrix} \alpha_{11} & \alpha_{12} \\ \alpha_{21} & \alpha_{22} \end{pmatrix}.$$

  If $\mathcal{E}$ had infinite rank (e.g., $\mathcal{E} = \bigoplus_{k=1}^{\infty} \mathcal{O}_{\mathcal{M}}$), one would need infinite matrices, and the tensor product $\mathcal{E}^\vee \otimes \mathcal{E}$ would not equal $\mathrm{End}\,(\mathcal{E})$.
- **Dual Basis**: The pairing $\langle \mathbf{e}_k, \mathbf{e}^\ell \rangle = \delta_k^\ell$ requires a **basis**. In our example:

$$\mathcal{E}^\vee \otimes \mathcal{E} = \langle \mathrm{d}x, \mathrm{d}y \rangle \otimes \langle \tfrac{\partial}{\partial x}, \tfrac{\partial}{\partial y} \rangle$$

  has basis elements $\mathrm{d}x \otimes \frac{\partial}{\partial x}$, $\mathrm{d}x \otimes \frac{\partial}{\partial y}$, etc. The isomorphism:

$$\mathrm{d}x \otimes \tfrac{\partial}{\partial x} \mapsto \left(\tfrac{\partial}{\partial x} \mapsto \langle \tfrac{\partial}{\partial x}, \mathrm{d}x \rangle \tfrac{\partial}{\partial x} = \tfrac{\partial}{\partial x}\right)$$

  relies on the dual basis existing locally. If $\mathcal{E}$ weren't locally free (e.g., a skyscraper sheaf), no such basis would exist.
- **Matrix Representation**: The correspondence:

$$\mathrm{Hom}_{\mathcal{O}_{\mathcal{M}}}(\mathcal{E}, \mathcal{E} \otimes \Omega^2_{\mathcal{M}}) \leftrightarrow n \times n \text{ matrices with entries in } \Omega^2_{\mathcal{M}}$$

  requires finite n. In our example, ϕ corresponds to:

$$\begin{pmatrix} \alpha_{11}\mathrm{d}x \wedge \mathrm{d}y & \alpha_{12}\mathrm{d}x \wedge \mathrm{d}y \\ \alpha_{21}\mathrm{d}x \wedge \mathrm{d}y & \alpha_{22}\mathrm{d}x \wedge \mathrm{d}y \end{pmatrix} \in \mathrm{End}\,(\mathcal{E}) \otimes \Omega^2_{\mathcal{M}}.$$

Now if the rank is infinite, the matrix multiplication is not well-defined for arbitrary sections.

△ Counterexample

> The following counterexample is presented as a guided exercise for readers interested in working through all the details of the proof and in understanding the deeper mechanism behind it.

The counterexample aims to demonstrate that the isomorphism fails when $\mathcal{E}$ is not locally free of finite rank. Let $\mathcal{M} = \mathbb{R}$, $\mathcal{I} =$ ideal sheaf of $\{0\}$. Locally,

$$\mathcal{I}(U) = \begin{cases} \{f \in C^\infty(U) \mid f(0) = 0\}, & \text{if } 0 \in U, \\ C^\infty(U), & \text{otherwise.} \end{cases}$$

We check now that $\mathcal{I}$ is **not locally free**. At $0, \mathcal{I}_0 =$ {germs of smooth functions vanishing at 0} $\subset \mathcal{O}_{\mathbb{R},0}$. $\mathcal{I}_0$ is the maximal ideal $\mathfrak{m}_0$ in the local ring $\mathcal{O}_{\mathbb{R},0}$. However, $\mathfrak{m}_0$ is not principal (find a proof of this!). Thus, $\mathcal{I}_0$ is not a free $\mathcal{O}_{\mathbb{R},0}$-module, and therefore, $\mathcal{I}$ is not locally free at 0. Consider:

$$\mathrm{Hom}_{\mathcal{O}_\mathbb{R}}(\mathcal{I}, \mathcal{I} \otimes \Omega^1_\mathbb{R})$$

Sections of $\mathcal{I}$ vanish at 0: for $f \in \mathcal{I}(U)$, $f(0) = 0$ if $0 \in U$. Morphisms must preserve vanishing: Any $\phi \in \mathrm{Hom}(\mathcal{I}, \mathcal{I} \otimes \Omega^1)$ must satisfy $\phi(f)(0) = 0$ for $f \in \mathcal{I}$. We analyze now the following:

$$\mathrm{End}\,(\mathcal{I}) \otimes \Omega^1_\mathbb{R} \cong \mathrm{Hom}(\mathcal{I}, \mathcal{I}) \otimes \Omega^1_\mathbb{R}.$$

By the endomorphism sheaf, we have $\mathrm{End}\,(\mathcal{I}) = \mathrm{Hom}_{\mathcal{O}_\mathbb{R}}(\mathcal{I}, \mathcal{I})$. Furthermore, there exists a global section $\mathrm{id} \in \Gamma(\mathbb{R}, \mathrm{End}\,(\mathcal{I}))$ (identity morphism).

Consider elements like $\mathrm{id} \otimes \mathrm{d}x$, where $(\mathrm{id} \otimes \mathrm{d}x)(f) = f \otimes \mathrm{d}x$, where $f(0)$ vanishes at 0. The isomorphism would imply a natural map:

$$\Phi : \mathrm{End}(\mathcal{I}) \otimes \Omega^1_\mathbb{R} \to \mathrm{Hom}_{\mathcal{O}_\mathbb{R}}\left(\mathcal{I}, \mathcal{I} \otimes \Omega^1_\mathbb{R}\right),$$

defined by

$$\Phi(\alpha \otimes \omega)(f) = \alpha(f) \otimes \omega,$$

where $\alpha \in \mathrm{End}\,(\mathcal{I})$ (an endomorphism of the ideal sheaf $\mathcal{I}$, so $\alpha : \mathcal{I} \to \mathcal{I}$ is $\mathcal{O}_\mathbb{R}$-linear) and $\omega \in \Omega^1_\mathbb{R}$. The element $\alpha \otimes \omega$ is a section of the tensor product sheaf $\mathrm{End}\,(\mathcal{I}) \otimes \Omega^1_\mathbb{R}$.

The output $\Phi(\alpha \otimes \omega)$ is an $\mathcal{O}_\mathbb{R}$-linear morphism $\mathcal{I} \to \mathcal{I} \otimes \Omega^1_\mathbb{R}$, and for a section $f \in \mathcal{I}(U)$, the result is $\alpha(f) \otimes \omega$.

In fact, Φ arises from the the universal property of tensor products. Elements of $\mathrm{End}\,(\mathcal{I}) \otimes \Omega^1_\mathbb{R}$ should induce $\mathcal{O}_\mathbb{R}$-linear maps $\mathcal{I} \to \mathcal{I} \otimes \Omega^1_\mathbb{R}$ (via evaluation).

However, the crucial point is that the map

$$\Phi : \mathrm{End}\,(\mathcal{I}) \otimes \Omega^1_{\mathbb{R}} \longrightarrow \mathrm{Hom}_{\mathcal{O}_{\mathbb{R}}}(\mathcal{I}, \mathcal{I} \otimes \Omega^1_{\mathbb{R}})$$

fails to be surjective. We now justify this claim by constructing an explicit morphism that does not lie in the image of Φ.

Consider the $\mathcal{O}_{\mathbb{R}}$-module homomorphism

$$\psi \in \mathrm{Hom}_{\mathcal{O}_{\mathbb{R}}}(\mathcal{I}, \mathcal{I} \otimes \Omega^1_{\mathbb{R}}), \qquad \psi(f) = x\, f' \otimes \mathrm{d}x,$$

where x denotes the coordinate function on $\mathbb{R}$ and $f' = \frac{\mathrm{d}f}{\mathrm{d}x}$ is the derivative of f. Note that $x\, f'$ vanishes at 0, hence indeed $\psi(f) \in \mathcal{I} \otimes \Omega^1_{\mathbb{R}}$ for all $f \in \mathcal{I}$. Furthermore, the Leibniz rule ensures that ψ is $\mathcal{O}_{\mathbb{R}}$-linear.

Assume, for the sake of contradiction, that ψ lies in the image of Φ. Then there would exist an element $\alpha \otimes \omega \in \mathrm{End}\,(\mathcal{I}) \otimes \Omega^1_{\mathbb{R}}$ such that

$$\Phi(\alpha \otimes \omega)(f) = \psi(f) \qquad \text{for all } f \in \mathcal{I}.$$

Unwinding the definition of Φ gives

$$\alpha(f) \otimes \omega = xf' \otimes \mathrm{d}x \qquad \text{for all } f,$$

which forces $\omega = \mathrm{d}x$ and $\alpha(f) = xf'$.

Here lies the contradiction: The map $f \mapsto xf'$ depends on the derivative of f and therefore is not an $\mathcal{O}_{\mathbb{R}}$-linear endomorphism of $\mathcal{I}$ arising from pointwise multiplication. In contrast, elements of the form $\Phi(\alpha \otimes \omega)$ are determined purely by pointwise evaluations of f, with no reference to its derivatives.

Consequently, $\psi \notin \mathrm{Im}(\Phi)$, and we conclude that Φ is not surjective.

Therefore, ψ is not in the image of Φ. Hence, the isomorphism fails and this ends the proof.

Remark 6.5 The reader can check the above proof on examples for $f(x) = x$ and $f(x) = x^2$.

6.3.4 Sectional Curvature

In this subsection we go back to our initial differential geometric language.

Let $\mathcal{M} \subset \mathbb{R}^n$ be a smooth m-dimensional submanifold. Let $p \in \mathcal{M}$ and let $E \subset \mathcal{T}(\mathcal{M})$ be a two-dimensional linear subspace of the tangent space. The sectional curvature of $\mathcal{M}$ at (p, E) is the number

$$K(p, E) = \frac{\langle R_p(u, v)v, u\rangle}{|u|^2|v|^2 - \langle u, v\rangle^2},$$

where $u, v \in E$ are linearly independent and R_p is the Riemannian curvature tensor.

6.3.4.1 Discussion Around This Concept

The Riemann curvature tensor R_p acts as an endomorphism:

$$R_p(u, v) : \mathcal{T}_p(\mathcal{M}) \to \mathcal{T}_p(\mathcal{M}), \quad w \mapsto R_p(u, v)w.$$

Thus $R_p(u, v)v \in \mathcal{T}_p(\mathcal{M})$, and $\langle R_p(u, v)v, u\rangle$ is a scalar. This is the standard coordinate-free expression for the sectional curvature numerator.

Denominator $|u|^2|v|^2 - \langle u, v\rangle^2$

This equals the squared area of the parallelogram spanned by u and v:

$$|u \wedge v|^2 = |u|^2|v|^2 - \langle u, v\rangle^2$$

This normalization ensures that $K(p, E)$ depends only on the 2-plane E, not the choice of basis vectors.

Basis Invariance

Let $\{\tilde{u}, \tilde{v}\}$ be another basis for E:

$$\tilde{u} = au + bv, \quad \tilde{v} = cu + dv, \quad \Delta = ad - bc \neq 0.$$

The numerator transforms as:

$$\langle R_p(\tilde{u}, \tilde{v})\tilde{v}, \tilde{u}\rangle = \Delta^2 \langle R_p(u, v)v, u\rangle.$$

The denominator transforms as:

$$|\tilde{u}|^2|\tilde{v}|^2 - \langle \tilde{u}, \tilde{v}\rangle^2 = \Delta^2(|u|^2|v|^2 - \langle u, v\rangle^2).$$

Thus, the ratio $K(p, E)$ is invariant:

$$\frac{\langle R_p(\tilde{u}, \tilde{v})\tilde{v}, \tilde{u}\rangle}{|\tilde{u}|^2|\tilde{v}|^2 - \langle \tilde{u}, \tilde{v}\rangle^2} = \frac{\langle R_p(u, v)v, u\rangle}{|u|^2|v|^2 - \langle u, v\rangle^2}.$$

Geometric Interpretation

- $K(p, E)$ generalizes Gaussian curvature to higher dimensions.
- For surfaces ($m = 2$), $K(p, \mathcal{T}_p(\mathcal{M}))$ equals Gaussian curvature.
- Measures intrinsic curvature of $\mathcal{M}$ along plane E at p.

Example: Unit Sphere $S^2 \subset \mathbb{R}^3$

At $p = (0, 0, 1)$ with $\mathcal{T}_p(S^2) = \text{span}\{e_1, e_2\}$:

$$R_p(e_1, e_2)e_2 = e_1, \quad \langle R_p(e_1, e_2)e_2, e_1\rangle = 1$$

$$|e_1|^2|e_2|^2 - \langle e_1, e_2\rangle^2 = 1 - 0 = 1$$

$$K(p, \mathcal{T}_p(S^2)) = 1/1 = 1 \quad \text{(matches Gaussian curvature of } S^2\text{)}.$$

Exercise 6.6 Let us consider the manifold of symmetric positive definite matrices, i.e.,

$$\{A \in \mathrm{Mat}_{n\times n}(\mathbb{R}) \mid A^{\mathrm{T}} = A,\ x^{\mathrm{T}}Ax > 0\}.$$

Compute the sectional curvature for this manifold.

Remark 6.6 The space of symmetric positive definite matrices has many interesting applications.

- Optimisation (convex programming).
- Information geometry and harmonic analysis (via Wishart laws). These are important in random matrix theory and statistics.
- Medicine (diffusion tensor imaging) as a model of the anisotropic diffusion of water in the tissues.
- Machine Learning: One is interested in finding the correlation between given properties (such as the weight/volume or weight and height in a given population). One models the relationship with a covariance matrix.

Definition 6.10 Let $k \in \mathbb{R}$ and $m \geq 2$ be an integer. An m-manifold has constant sectional curvature k if and only if $K(p, E) = k$ for every $p \in \mathcal{M}$ and every two-dimensional linear subspace $E \subset \mathcal{T}_p\mathcal{M}$.

Theorem 6.1 *Let $\mathcal{M}$ be an m-manifold. Fix an element $p \in \mathcal{M}$ and a real number k. Then the following are equivalent: $K(p, E) = k$ for every two-dimensional linear subspace $E \subset \mathcal{T}_p\mathcal{M}$. The Riemann curvature tensor of $\mathcal{M}$ at p is given by:*

$$\langle R_p(v_1, v_2)v_3, v_4\rangle = k(\langle v_1, v_4\rangle - \langle v_2, v_3\rangle),$$

for all $v_1, \ldots, v_4 \in \mathcal{T}_p\mathcal{M}$.

6.4 Torsion in Differential Geometry

6.4.1 Torsion Tensor

Let $\mathcal{M}$ be a smooth manifold, and let ∇ be an affine connection on $\mathcal{M}$. The presence of torsion in ∇ quantifies the extent to which infinitesimal parallel transport fails to be symmetric. Intuitively, the torsion of ∇ quantifies the failure of infinitesimal parallelograms formed by vector flows to close.

Definition 6.11 The *torsion tensor* T associated with a connection ∇ is the $(1, 2)$-tensor defined by

$$\mathrm{T}(X, Y) = \nabla_X(Y) - \nabla_Y(X) - [X, Y],$$

for any vector fields X, Y on $\mathcal{M}$, where $[X, Y]$ denotes the Lie bracket and $\nabla_Y(X)$ is the covariant derivative. The connection ∇ is said to be *torsion-free* if $\mathrm{T} = 0$.

6.4.2 Local Coordinates

In local coordinates, $\{x^i\}$, the torsion tensor can be expressed in terms of the connection coefficients Γ^k_{ij}:

$$\mathrm{T}^i_{jk} = \Gamma^i_{jk} - \Gamma^i_{kj},$$

where T^i_{jk} are the components of the torsion tensor, and Γ^i_{jk} are the *connection coefficients*.[5]

- The torsion measures the failure of flows of X and Y to form a closed infinitesimal parallelogram. For small ε, the quadrilateral formed by flowing ε along X, ε along Y, $-\varepsilon$ along X, $-\varepsilon$ along Y fails to close by $\varepsilon^2\mathrm{T}(X, Y) + \mathcal{O}(\varepsilon^3)$.
- Let us mention that the torsion affects the *holonomy* of a connection, which describes how vectors rotate when parallel transported around closed loops. The torsion contributes to the *affine part* of holonomy. When parallel-transporting vectors along closed loops, torsion causes translation in addition to rotation (which comes from curvature).
- The geometric interpretation of torsion as "parallelogram non-closure" inherently implies the connection is asymmetric, i.e., $\Gamma^i_{jk} \neq \Gamma^i_{kj}$.

In Riemannian geometry, the Levi–Civita connection is the unique torsion-free connection that is compatible with the metric. This connection is central to the study of curvature and geodesics. The connection coefficients Γ^i_{jk} are referred to as *Christoffel symbols* only in the special case of the Levi–Civita connection. For more general connections—such as those on manifolds without a metric, or metric connections that allow torsion—Γ^i_{jk} should simply be called *connection coefficients*, not Christoffel symbols.

[5] There is a subtle distinction between the notion of connection coefficients and of Christoffel symbols.

6.4.3 Geodesic Deviations

6.4.3.1 Geodesic Deviation with Torsion

In the presence of torsion, the equation of geodesic deviation (describing how nearby geodesics spread apart or converge) includes additional terms involving the torsion tensor.

Intuitively, imagine two nearby geodesics on a surface. If the surface is flat (e.g., a plane), geodesics are straight lines with constant separation. On a curved surface (e.g., a sphere), geodesics converge or diverge. Torsion introduces *twisting* to this deviation.

Let $\mathcal{M}$ be a smooth manifold equipped with:

- the sheaf $\mathcal{O}_{\mathcal{M}}$ of smooth functions;
- the sheaf $\mathcal{E} = \mathcal{T}(\mathcal{M})$ of tangent bundle sections;
- an affine connection ∇ on $\mathcal{E}$ with torsion T.

A *geodesic* is a path γ satisfying:

$$\nabla_{\dot{\gamma}}\dot{\gamma} = 0.$$

Consider a one-parameter family of geodesics:

$$\wp : I \times (-\epsilon, \epsilon) \to \mathcal{M},$$

with $I \subset \mathbb{R}$ parametrizing the curve. The *geodesic deviation vector field* is:

$$\mathcal{J} = \left.\frac{\partial \wp}{\partial s}\right|_{s=0} \in \Gamma(\gamma^*\mathcal{E}).$$

This satisfies the *generalized Jacobi equation*:

$$\nabla_{\dot{\gamma}}\nabla_{\dot{\gamma}}\mathcal{J} + R(\dot{\gamma}, \mathcal{J})\dot{\gamma} = \nabla_{\dot{\gamma}}\left(\mathrm{T}(\mathcal{J}, \dot{\gamma})\right) + \mathrm{T}(\nabla_{\dot{\gamma}}\mathcal{J}, \dot{\gamma}),$$

where $R \in \Gamma(\mathrm{End}(\mathcal{E}) \otimes \Omega^2_{\mathcal{M}})$ is the curvature tensor of ∇. For torsion-free connections ($\mathrm{T} = 0$), this reduces to the usual Jacobi equation.

Remark 6.7 In the following part of the book on the geometry of information, we focus on the case where the torsion vanishes.

- **Key Insight**: Torsion terms $\nabla_{\dot{\gamma}}(\mathrm{T}(\mathcal{J}, \dot{\gamma}))$ and $\mathrm{T}(\nabla_{\dot{\gamma}}\mathcal{J}, \dot{\gamma})$ describe how the connection's asymmetry affects geodesic separation.
- **Riemannian Case**: For Levi–Civita connections ($\mathrm{T} = 0$), the standard Jacobi equation governs deviation purely via curvature.

6.4.3.2 Flat Versus Curved Geometry

In flat geometry (that is when $\mathcal{M}$ is locally isometric to $\mathbb{R}^n$ with Euclidean metric), the fact that $R = 0$ reduces the Jacobi equation to:

$$\nabla_{\dot{\gamma}} \nabla_{\dot{\gamma}} \mathcal{J} = 0.$$

This implies $\mathcal{J}$ evolves *linearly* along γ, meaning that:

- *Parallel* geodesics remain equidistant.
- *Non-parallel* geodesics diverge linearly at constant rate.

In curved geometry, the curvature term $R(\dot{\gamma}, \mathcal{J})\dot{\gamma}$ governs relative acceleration:

- $\langle R(\dot{\gamma}, \mathcal{J})\dot{\gamma}, \mathcal{J} \rangle > 0 \Rightarrow$ geodesics converge (positive curvature).
- $\langle R(\dot{\gamma}, \mathcal{J})\dot{\gamma}, \mathcal{J} \rangle < 0 \Rightarrow$ geodesics diverge (negative curvature).
- **Off-diagonal components** (orthogonal to $\mathcal{J}$) induce *shearing*: geodesics slide past each other while maintaining constant distance.

Thus, geodesic deviation provides an $\mathcal{O}_{\mathcal{M}}$-linear measure of curvature's local obstruction to Euclidean geometry.

6.4.4 Torsion Tensor

The torsion tensor is *skew-symmetric* in its vector field arguments:

$$\mathrm{T}(X, Y) = -\mathrm{T}(Y, X) \quad \forall \text{ vector fields } X, Y.$$

In local coordinates, this implies ${\mathrm{T}^i}_{jk} = -{\mathrm{T}^i}_{kj}$ for its components.

If $\mathrm{T} = 0$, the connection is torsion-free. In local coordinates, the connection coefficients satisfy $\Gamma^i_{jk} = \Gamma^i_{kj}$, making the connection *symmetric in its lower indices*.

6.4.5 Relations to Curvature

The torsion and curvature tensors satisfy the **First Bianchi Identity**:

$$\begin{aligned} &R^{\nabla}(X, Y)Z + R^{\nabla}(Y, Z)X + R^{\nabla}(Z, X)Y \\ &\quad = (\nabla_X \mathrm{T})(Y, Z) + (\nabla_Y \mathrm{T})(Z, X) + (\nabla_Z \mathrm{T})(X, Y) \\ &\qquad + \mathrm{T}(\mathrm{T}(X, Y), Z) + \mathrm{T}(\mathrm{T}(Y, Z), X) + \mathrm{T}(\mathrm{T}(Z, X), Y). \end{aligned} \tag{6.30}$$

The identity arises from the Jacobi identity for the Lie bracket of vector fields and the definitions of curvature and torsion.

When $\mathrm{T} = 0$, it reduces to the classical Riemannian Bianchi identity.

When $\mathrm{T} \neq 0$, the failure of the Jacobi identity introduces:

- first-order derivatives of T (coming from $\nabla\mathrm{T}$);
- algebraic products $\mathrm{T} \circ \mathrm{T}$ (arising from the non-closure of infinitesimal parallelograms).

In components, this becomes:

$$R^{\lambda}{}_{[\mu\nu\rho]} = \nabla_{[\mu}\mathrm{T}^{\lambda}{}_{\nu\rho]} + \mathrm{T}^{\sigma}{}_{[\mu\nu}\mathrm{T}^{\lambda}{}_{\rho]\sigma}$$

where $[\mu\nu\rho]$ denotes antisymmetrization[6] over indices and where:

$$\nabla_{\mu}\mathrm{T}_{\nu\rho}{}^{\lambda}$$

is linear in T and its derivatives, while

$$\mathrm{T}_{\mu\nu}{}^{\sigma}\mathrm{T}_{\rho\sigma}{}^{\lambda}$$

is quadratic in T. Notably, there are no terms of the form T^3, $(\nabla\mathrm{T})\mathrm{T}$, or $\nabla^2\mathrm{T}$.

For a categorical perspective, consider the *Atiyah algebroid* $A_{\mathcal{E}}$ associated to a vector bundle $\mathcal{E}$:

$$0 \to \mathrm{End}(\mathcal{E}) \to A_{\mathcal{E}} \xrightarrow{\sigma} T(\mathcal{M}) \to 0.$$

[6] For any tensor $A_{\mu\nu\rho}$, the antisymmetrization version is

$$A_{[\mu\nu\rho]} = \frac{1}{3!}\sum_{\sigma\in S_3} \mathrm{sgn}(\sigma)\, A_{\sigma(\mu)\sigma(\nu)\sigma(\rho)}.$$

where S_3 is the permutation group of three elements and $\mathrm{sgn}(\sigma)$ is the sign of the permutation. For instance for the term $\nabla_{[\mu}\mathrm{T}_{\nu\rho]}^{\lambda}$, the antisymmetrization of the covariant derivative of torsion is:

$$\nabla_{[\mu}\mathrm{T}_{\nu\rho]}^{\lambda} = \frac{1}{3}\left(\nabla_{\mu}\mathrm{T}_{\nu\rho}^{\lambda} + \nabla_{\nu}\mathrm{T}_{\rho\mu}^{\lambda} + \nabla_{\rho}\mathrm{T}_{\mu\nu}^{\lambda}\right).$$

For the second term $\mathrm{T}_{[\mu\nu}{}^{\sigma}\mathrm{T}_{\rho]\sigma}^{\lambda}$. This combines two parts: (i) Antisymmetrization of $\mathrm{T}_{\mu\nu}^{\sigma}$ over μ, ν, ρ (since $[\,\mu\nu\rho\,]$ applies to all three indices).

(ii) Contraction with $\mathrm{T}_{\rho\sigma}^{\lambda}$.

Explicitly,

$$\mathrm{T}_{[\mu\nu}{}^{\sigma}\mathrm{T}_{\rho]\sigma}^{\lambda} = \frac{1}{3}\left(\mathrm{T}_{\mu\nu}^{\sigma}\mathrm{T}_{\rho\sigma}^{\lambda} + \mathrm{T}_{\nu\rho}^{\sigma}\mathrm{T}_{\mu\sigma}^{\lambda} + \mathrm{T}_{\rho\mu}^{\sigma}\mathrm{T}_{\nu\sigma}^{\lambda}\right).$$

A connection ∇ corresponds to a splitting σ^{-1}. In this framework:

- Curvature R^∇ is the obstruction to ∇ being a Lie algebroid morphism.
- Torsion T measures the failure of ∇ to preserve Lie brackets when $\mathcal{E} = \mathcal{T}(\mathcal{M})$.

6.4.6 Spencer Differential and Bianchi Identity

Let E be a vector bundle over $\mathcal{M}$ and let α be an E-valued k-form. The **Spencer differential** for a connection ∇ on a vector bundle E is the operator:

$$d^\nabla : \Gamma(\wedge^k \mathcal{T}^*\mathcal{M} \otimes E) \to \Gamma(\wedge^{k+1} \mathcal{T}^*\mathcal{M} \otimes E)$$

defined for $\alpha \in \Gamma(\wedge^k \mathcal{T}^*\mathcal{M} \otimes E)$ and vector fields $X_0, \ldots, X_k$ by:

$$\begin{aligned}(\mathrm{d}^\nabla \alpha)(X_0, \ldots, X_k) &= \sum_{i=0}^{k} (-1)^i \nabla_{X_i} \alpha(X_0, \ldots, \widehat{X_i}, \ldots, X_k) \\ &+ \sum_{0 \le i < j \le k} (-1)^{i+j} \alpha([X_i, X_j], X_0, \ldots, \widehat{X_i}, \ldots, \widehat{X_j}, \ldots, X_k).\end{aligned}$$

For the tangent bundle $E = \mathcal{T}\mathcal{M}$ with torsion $\mathrm{T} \in \Gamma(\wedge^2 \mathcal{T}^*\mathcal{M} \otimes \mathcal{T}\mathcal{M})$, the first Bianchi identity is:

$$\mathrm{d}^\nabla \mathrm{T} = \mathfrak{S}(R^\nabla) - \mathfrak{S}(\mathrm{T} \circ \mathrm{T}),$$

where:

- $\mathfrak{S}$ denotes cyclic summation: $\mathfrak{S}(A)(X, Y, Z) = A(X, Y, Z) + A(Y, Z, X) + A(Z, X, Y)$.
- $\mathrm{T} \circ \mathrm{T}(X, Y, Z) = \mathrm{T}(\mathrm{T}(X, Y), Z) +$ cyclic permutations.

Geometric Iinterpretation. When parallel transporting along paths:

- The failure of $d^\nabla \mathrm{T} = 0$ reflects path-dependence in transporting torsion.
- The identity encodes how torsion T and curvature R^∇ constrain each other.

Example 6.2 The torsion can be interpreted as an obstruction to the integrability of certain geometric structures, such as foliations or distributions on a manifold.

Example 6.3 The torsion plays a central role in non-Riemannian geometries. This includes Finsler geometry for instance and teleparallel gravity. In those frameworks, the torsion tensor is used to describe the gravitational field, replacing the curvature tensor of general relativity.

Part II

Probabilities, Statistics, and Related Topics

Probability Theory—Measure Theory—Statistics 7

In this chapter, we give a short overview on probabilistic and statistic notions. We will see that everything that has seen introduced so far can be used in the context of probability and statistics, namely information theory. Such a mixture gave birth to the theory of information geometry.

We will also recall some notions of distance or divergence between probabilities, which is important for the "learning process".

7.1 Key Information for Statistics

When studying a mathematical problem in statistics, several key pieces of information come into play:

1. **Qualitative Description of the Observed Phenomenon**:
   This refers to a detailed understanding of the nature and characteristics of the event or situation being observed, which can be represented through various inputs ω_i collected from the sample space Ω.
2. **Information About the Unknown Distribution of Outcomes** ω:
   This involves understanding how different outcomes are distributed, which is fundamental to statistical analysis.
3. **A Priori Probabilities**:
   This entails the probabilities associated with observing various phenomena, represented as $(\Omega, \mathbf{S}, P_\theta)$. Here, $\mathbf{S}$ denotes a σ-algebra that organizes measurable sets for the outcomes.

© The Author(s), under exclusive license to Springer Nature Switzerland AG 2026
N. C. Combe et al., *Exploring Information Geometry*, Frontiers in Mathematics,
https://doi.org/10.1007/978-3-032-23054-6_7

4. **A Distribution on the Parameter Space** $Q\{d\theta\}$:
   This refers to a probability distribution defined on the range of possible parameter values that could explain the outcomes, denoted by the parameter space Θ.

In this context, the first item is associated with a collection of inputs gathered from the sample space Ω. The second item involves the probability distribution $P[\cdot]$, which comes from a family of distributions P_θ, where the parameter θ belongs to the parameter space Θ. Overall, this transforms the issue of mathematical statistics into a decision-making problem, as one must determine which pieces of information are most beneficial. Formally, every statistical problem relates to a measurable space $(\Omega, \mathbf{S})$, which captures the outcomes of the observed phenomenon, a family of probability distributions P_θ, and potentially additional a priori knowledge about the unknown phenomenon linked to another measurable space $(\Omega', \mathbf{S}')$, which may concern hypotheses or actions taken.

7.2 Probability

7.2.1 Probability Space

The foundational concept in probability theory is that of a *random experiment*. This is mathematically modeled by a *probability space*, defined as the triplet $(\Omega, \mathbf{S}, P)$ where:

1. Ω is the *sample space*: the set of all possible outcomes (or sample points) ω of the experiment.
2. $\mathbf{S}$ is a $\boldsymbol{\sigma}$*-algebra* of subsets of Ω, called the *family of events*. Every event $A \in \mathbf{S}$ corresponds to a measurable subset of Ω.
3. P is a *probability measure*, a function $P : \mathbf{S} \to [0, 1]$ assigning probabilities to events (satisfying the axioms below).

Definition 7.1 **(σ-algebra)** A collection $\mathbf{S}$ of subsets of Ω is a σ-algebra if it satisfies:

1. $\Omega \in \mathbf{S}$;
2. $A \in \mathbf{S} \implies A^c \in \mathbf{S}$ (closed under complements);
3. $A_1, A_2, \ldots \in \mathbf{S} \implies \bigcup_{i=1}^{\infty} A_i \in \mathbf{S}$ (closed under countable unions).

The pair $(\Omega, \mathbf{S})$ is called a **measurable space**.

Definition 7.2 (Probability measure) A function $P : \mathbf{S} \to [0, 1]$ is a probability measure if:

1. $P(\Omega) = 1$ (normalization),
2. For any countable sequence $\{A_i\}$ of pairwise disjoint events in $\mathbf{S}$,

$$P\left(\bigcup_{i=1}^{\infty} A_i\right) = \sum_{i=1}^{\infty} P(A_i) \qquad \text{(countable additivity).}$$

The triplet $(\Omega, \mathbf{S}, P)$ with these properties is a **probability space**.

Example 7.1 • The *smallest* σ-algebra containing one $A \subset \Omega$ is $\mathbf{S} = \{\emptyset, A, A^c, \Omega\}$.

- The *largest* σ-algebra is consisted of all subset of Ω. This algebra served us in discrete spaces, but is too large to be useful in general.
- If Ω is a topological space the σ-algebra generated by all the open subset of Ω is called the *Borel's algebra.*

Exercise 7.1 Show that the σ-algebra generated by n disjoint subsets of a set Ω has cardinality 2^n.

Definition 7.3 (Probability measure) A **probability** P is a function

$$P : \mathbf{S} \to [0, 1]$$

satisfying the following axioms:

1. $P[\Omega] = 1$ (normalization)
2. For any countable sequence $\{A_i\}$ of pairwise disjoint events in $\mathbf{S}$ ($A_i \cap A_j = \emptyset$ for $i \neq j$):

$$P\left[\bigcup_{i=1}^{\infty} A_i\right] = \sum_{i=1}^{\infty} P[A_i]$$

(countable additivity)

- The triplet $(\Omega, \mathbf{S}, P)$ is called a **probability space**.
- An event $A \in \mathbf{S}$ with $P[A] = 0$ is called a P**-null set**.
- An event $A \in \mathbf{S}$ with $P[A] = 1$ is said to occur **almost surely** (denoted "a.s.").

From these axioms, we derive these fundamental relations:

- $P[A^c] = 1 - P[A]$, for $A \in \mathbf{S}$
- $P[A \setminus B] = P[A] - P[B]$, for $B \subseteq A$ with $A, B \in \mathbf{S}$ (implies $P[B] \leq P[A]$ when $B \subseteq A$)
- $P[A \cup B] = P[A] + P[B] - P[A \cap B]$, for $A, B \in \mathbf{S}$
- $P[\emptyset] = 0$ (derivable from the axioms).

7.3 Examples

To illustrate the construction of probability spaces, let us discuss the following examples.

7.3.1 Flipping Coin Examples: How Biased Is It?

Example 7.2 A coin is flipped independently N times. The sequence of outcomes $(\omega_1, \ldots, \omega_N)$, consisting of heads H and tails T, might appear as

$$\underbrace{HTHTHH\cdots THT}_{N \text{ symbols}}.$$

To determine whether the coin is biased, we analyze the empirical frequency of heads.

The fundamental question is: Can we assume the coin is fair? Equivalently: Does the true probability p of heads satisfy $|p - \frac{1}{2}| < 10^{-3}$?

Now consider a more structured probabilistic model:

Example 7.3 The coin is flipped independently N times, with unknown heads probability θ. We assume the probability of any outcome sequence ω follows the Bernoulli model:

$$P(\omega) = \theta^{k(\omega)}(1 - \theta)^{N-k(\omega)}$$

where $k(\omega)$ counts heads in ω.

This defines a probability distribution on Ω (the sample space of 2^N possible sequences). Explicitly, for sequences indexed by $j = 1, \ldots, 2^N$:

$$P(\{\omega_j\}) = \theta^{k(\omega_j)}(1 - \theta)^{N-k(\omega_j)}, \quad \theta \in [0, 1].$$

Geometrically, this parameterization traces a curve in the $(2^N - 1)$-dimensional simplex of probability distributions on Ω.

Under lexicographical ordering (with $T \prec H$), the sequence ω_j corresponds to the binary expansion of $j-1$ (padded to N bits), where:

- T encodes 0, H encodes 1
- $k(\omega_j)$ = number of 1's in binary $(j-1)$.

To conclude our coin flip example, we observe:

Example 7.4 • For $N=1$: The sample space has $2^1=2$ outcomes (H and T). The probability distribution is parameterized by the curve:

$$P(\{\omega_j\}) = \theta^{k(\omega_j)}(1-\theta)^{1-k(\omega_j)}, \quad j=1,2,\ \theta \in [0,1]$$

in the 1-dimensional probability simplex (a line segment between $[P(H), P(T)] = (1,0)$ and $(0,1)$).

- For $N=2$: The sample space has 4 outcomes (HH, HT, TH, TT). The distribution is parameterized by:

$$P(\{\omega_j\}) = \theta^{k(\omega_j)}(1-\theta)^{2-k(\omega_j)}, \quad j=1,\ldots,4,\ \theta \in [0,1]$$

This describes a curve in the 3-dimensional probability simplex (a tetrahedron).
Curve behavior:

- At $\theta=0$: $P(TT)=1$ (vertex $(0,0,0,1)$ in order (HH, HT, TH, TT));
- At $\theta=\frac{1}{2}$: Uniform distribution $P(HH)=P(HT)=P(TH)=P(TT)=\frac{1}{4}$ (center point);
- At $\theta=1$: $P(HH)=1$ (vertex $(1,0,0,0)$).

The curve connects opposite vertices through the simplex center.

This coin flip model generalizes to broader probability contexts:

Example 7.5 (Multinomial distribution) For a finite sample space $\Omega=\{\omega_1,\ldots,\omega_m\}$ with power set σ-algebra $\mathcal{P}(\Omega)$, probabilities are defined by:

$$p_k = P(\{\omega_k\}), \quad \text{where} \quad \sum_{k=1}^{m} p_k = 1,\ p_k \geq 0$$

The probability measure can be expressed as:

$$P = \sum_{k=1}^{m} p_k \delta_{\omega_k}$$

where the notation δ_{ω_m} represents the Dirac measure (or point mass) concentrated at the outcome ω_m. In measure theory and probability, this is defined as:

$$\delta_{\omega_m}(A) = \begin{cases} 1 & \text{if } \omega_m \in A, \\ 0 & \text{if } \omega_m \notin A. \end{cases}$$

Equivalently, using the constraint $\sum_{k=1}^{m} p_k = 1$:

$$P = \sum_{k=1}^{m-1} p_k \delta_{\omega_k} + \left(1 - \sum_{k=1}^{m-1} p_k\right) \delta_{\omega_m}$$

The parameter space is the $(m-1)$-dimensional probability simplex. **Note**: The coin flip model is a special case where $m = 2^N$ and probabilities follow the binomial structure.

7.3.2 Other Examples

Example 7.6 (Discrete countable space) For a countably infinite sample space $\Omega = \{\omega_1, \omega_2, \ldots\}$ with the σ-algebra $\mathbf{S} = \mathcal{P}(\Omega)$ (power set), the probability measure is determined by:

$$p_k = P(\{\omega_k\}), \quad \text{where} \quad \sum_{k=1}^{\infty} p_k = 1, \; p_k \geq 0 \tag{7.1}$$

The Poisson distribution is a classic example:

$$p_k = \mathrm{e}^{-\lambda} \frac{\lambda^k}{k!}, \quad k = 0, 1, 2, \ldots \tag{7.2}$$

Example 7.7 (Continuous distribution) For continuous distributions, take $\Omega = \mathbb{R}$ with the Borel σ-algebra $\mathcal{B}(\mathbb{R})$ generated by intervals $\{(-\infty, a] : a \in \mathbb{R}\}$. The probability is defined via a density function ρ satisfying:

$$P\big((-\infty, a]\big) = \int_{-\infty}^{a} \rho(x)\mathrm{d}x \tag{7.3}$$

with

$$\rho(x) \geq 0 \quad \text{and} \quad \int_{-\infty}^{\infty} \rho(x)\mathrm{d}x = 1 \tag{7.4}$$

The Gaussian (normal) distribution is fundamental:

$$\rho(x) = \frac{1}{\sigma\sqrt{2\pi}} e^{-\frac{(x-\mu)^2}{2\sigma^2}} \tag{7.5}$$

Exercise 7.2 Prove the following theorem on σ-Algebras and subcovers. Let $(X, \mathcal{A}, \mu)$ be a measure space, and let $\mathcal{B} \subseteq \mathcal{A}$ be a sub-σ-algebra of $\mathcal{A}$. If $\{A_n\}_{n\in\mathbb{N}}$ is a countable cover of X by sets in $\mathcal{A}$, then there exists a subcover in $\mathcal{B}$ if and only if the measure μ on $\mathcal{A}$ is uniquely determined by its restriction to $\mathcal{B}$.

7.4 Measure

While probability measures require total measure 1, general measures provide a flexible framework for quantifying spaces without normalization constraints. This generalization enables applications across diverse fields including geometric measure theory, functional analysis, and stochastic processes.

Definition 7.4 (Measure) A measure μ on a measurable space $(\Omega, \mathbf{S})$ is a function

$$\mu : \mathbf{S} \to [0, +\infty]$$

satisfying:

1. $\mu(\emptyset) = 0$.
2. Countable additivity: For any sequence $\{A_i\}_{i=1}^{\infty}$ of pairwise disjoint sets in $\mathbf{S}$,

$$\mu\left(\bigcup_{i=1}^{\infty} A_i\right) = \sum_{i=1}^{\infty} \mu(A_i)$$

- A measure is *finite* (or bounded) if $\mu(\Omega) < \infty$.
  Note: Probability measures are finite measures with $\mu(\Omega) = 1$.
- A measure is σ*-finite* if Ω can be decomposed as

$$\Omega = \bigcup_{k=1}^{\infty} B_k \quad \text{with} \quad \mu(B_k) < \infty \ \forall k$$

Remark 7.1 σ-finite measures need not be finite. The Lebesgue measure λ on $\mathbb{R}$ is σ-finite since

$$\mathbb{R} = \bigcup_{k=1}^{\infty} [-k, k] \quad \text{with} \quad \lambda([-k, k]) = 2k < \infty,$$

yet $\lambda(\mathbb{R}) = \infty$.

Exercise 7.3 If X is a metric space and A and B are two disjoint closed subsets of X, then there exists a continuous function $f(x)$ on X with properties:

(1) $0 \le f(x) \le 1$,

(2) $\begin{cases} f(x) = 0 & \text{for } x \in A, \\ f(x) = 1 & \text{for } x \in B. \end{cases}$

7.5 Measurable Function—Random Variables

7.5.1 Random Variables and Measurability

Consider a random experiment modeled by a probability space $(\Omega, \mathbf{S}, P)$. A random variable is a measurable function from Ω to a measurable space $(E, \mathcal{B})$ that preserves event structure.

Definition 7.5 (Measurable function) Let $(\Omega, \mathbf{S})$ and $(E, \mathcal{B})$ be measurable spaces. A function $X : \Omega \to E$ is *measurable* if for every $B \in \mathcal{B}$:

$$X^{-1}(B) = \{\omega \in \Omega : X(\omega) \in B\} \in \mathbf{S}$$

Definition 7.6 (Random variable) Let $(\Omega, \mathbf{S}, P)$ be a probability space and $(E, \mathcal{B})$ a measurable space. An E-valued *random variable* is an equivalence class of measurable functions where

$$X \sim Y \quad \text{if} \quad P(X \neq Y) = 0 \quad \text{(equal } P\text{-almost everywhere)}$$

.

- A *real random variable* is a measurable function $X : \Omega \to \mathbb{R}$ where $\mathbb{R}$ has the Borel σ-algebra $\mathcal{B}(\mathbb{R})$.
- An *extended real random variable* maps to $\overline{\mathbb{R}} = \mathbb{R} \cup \{-\infty, +\infty\}$ with the extended Borel σ-algebra.
- A *discrete random variable* takes values in a countable set $E \subseteq \mathbb{R}$ (finite or infinite).
- A *random vector* is a measurable function $X : \Omega \to \mathbb{R}^n$ for $n \ge 2$.

Definition 7.7 (Expectation) For a random variable X on $(\Omega, \mathbf{S}, P)$, the *expectation* (or mean) is the Lebesgue integral:

$$\mathbb{E}_P[X] = \int_\Omega X(\omega)\mathrm{d}P(\omega)$$

defined when $X \in L^1(\Omega, \mathbf{S}, P)$.

7.6 Dominating Measure

Definition 7.8 (Absolute continuity) Let μ and ν be measures on $(\Omega, \mathbf{S})$. We say μ is *absolutely continuous* with respect to ν (denoted $\mu \ll \nu$) if:

$$\nu(A) = 0 \implies \mu(A) = 0 \quad \text{for all } A \in \mathbf{S}$$

When $\mu \ll \nu$, ν is called a *dominating measure* for μ.

Remark 7.2 (Properties of absolute continuity) Absolute continuity satisfies:

- **Reflexivity**: $\mu \ll \mu$
- **Transitivity**: $\mu \ll \nu$ and $\nu \ll \lambda \implies \mu \ll \lambda$
- **Non-antisymmetry**: $\mu \ll \nu$ and $\nu \ll \mu$ do *not* imply $\mu = \nu$.

Definition 7.9 (Equivalent measures) Measures μ and ν are *equivalent* ($\mu \equiv \nu$) if $\mu \ll \nu$ and $\nu \ll \mu$.

△ Absolute continuity induces a partial ordering on equivalence classes of measures.

Example 7.8 Let $\{P_k\}$ be a finite or countable family of probability measures on $(\Omega, \mathbf{S})$. Define:

- **Finite case**: $P_0 = \frac{1}{n}\sum_{k=1}^{n} P_k$
- **Countable case**: $P_0 = \sum_{k=1}^{\infty} \frac{1}{2^k} P_k$.

Then $P_k \ll P_0$ for all k. Moreover, P_0 is a probability measure.

While absolute continuity and measure equivalence provide fundamental relationships between "non-negative" measures, many applications—such as in functional analysis, physics, and probability theory—require a broader framework that accommodates measures with varying signs. This leads naturally to the concept of signed measures, which extend the notion of measure by allowing real-valued set functions.

Definition 7.10 (Signed measure) A *signed measure* on $(\Omega, \mathbf{S})$ is a function $\mu : \mathbf{S} \to \mathbb{R} \cup \{+\infty\}$ or $\mathbb{R} \cup \{-\infty\}$ (but not both) satisfying:

1. $\mu(\emptyset) = 0$
2. For any pairwise disjoint $\{A_i\}_{i=1}^{\infty} \subset \mathbf{S}$:

$$\mu\left(\bigcup_{i=1}^{\infty} A_i\right) = \sum_{i=1}^{\infty} \mu(A_i)$$

with absolute convergence if the left side is finite.

Theorem 7.1 (Jordan decomposition) *Every signed measure μ decomposes uniquely as $\mu = \mu^+ - \mu^-$ where:*

- *μ^+, μ^- are (non-negative) measures.*
- *There exists a partition $\Omega = \boldsymbol{\Omega}^+ \cup \boldsymbol{\Omega}^-$ in two sets $\boldsymbol{\Omega}^+$ and $\boldsymbol{\Omega}^-$ with $\mu^-(\boldsymbol{\Omega}^+) = \mu^+(\boldsymbol{\Omega}^-) = 0$.*
- *$|\mu| = \mu^+ + \mu^-$ (total variation measure).*

Definition 7.11 (Absolute continuity for signed measures) For a signed measure μ on $(\Omega, \mathbf{S})$, we say μ is *absolutely continuous* with respect to the measure ν ($\mu \ll \nu$) if:

$$\nu(A) = 0 \implies |\mu|(A) = 0 \quad \forall A \in \mathbf{S},$$

where $|\mu|$ is the *total variation measure*, defined by:

$$|\mu|(A) = \sup\left\{\sum_{k=1}^{n} |\mu(E_k)| : \{E_k\}_{k=1}^{n} \text{ is a finite } \mathbf{S}\text{-partition of } A\right\}.$$

Theorem 7.2 (Radon–Nikodym theorem) *Let μ and ν be σ-finite measures on $(\Omega, \mathbf{S})$ with $\mu \ll \nu$. Then there exists a ν-measurable function $\rho : \Omega \to [0, +\infty)$, called the* Radon–Nikodym derivative *(denoted $\rho = \frac{d\mu}{d\nu}$), such that for every $A \in \mathbf{S}$:*

$$\mu(A) = \int_A \rho d\nu.$$

When $\mu = P$ is a probability measure and ν is a σ-finite measure with $P \ll \nu$, the Radon–Nikodym derivative $\rho = \frac{\mathrm{d}P}{\mathrm{d}\nu}$ is called the *probability density function* with respect to ν. It satisfies:

$$P(A) = \int_A \rho \mathrm{d}\nu, \quad \forall A \in \mathbf{S},$$

$$\int_\Omega \rho \mathrm{d}\nu = 1.$$

Conversely, if $\nu \ll P$ where P is a probability measure and ν is any measure, then P *dominates* ν (i.e., P serves as a dominating measure for ν).

Remark 7.3 The Radon–Nikodym theorem fails for signed measure ν.

7.7 Information Geometry

Consider a parametric family of probability densities with respect to a dominating measure μ on $(\Omega, \mathbf{S})$:

$$\mathscr{S} = \left\{ \rho_\theta \in L^1(\Omega, \mathbf{S}, \mu) : \rho_\theta > 0 \ \mu\text{-a.e.}, \int_\Omega \rho_\theta \mathrm{d}\mu = 1, \ \theta \in \Theta \subseteq \mathbb{R}^n \right\}$$

where Θ is the parameter space. We assume $\mathscr{S}$ forms a smooth manifold—the statistical manifold.

A fundamental class is the exponential family, which means that densities admit the canonical form:

$$\rho_\theta(\omega) = \exp\left(\sum_{i=1}^n \theta^i X_i(\omega) - \psi(\theta) \right)$$

with:

- $\theta = (\theta^1, \ldots, \theta^n)$: Natural parameters.
- $X = (X_1, \ldots, X_n)$: Sufficient statistics.
- $\psi(\theta)$: Log-partition function (potential).

More generally, exponential families have the unified representation:

$$\rho_\theta(\omega) = h(\omega) \exp\left(\eta(\theta) \cdot X(\omega) - A(\theta)\right)$$

where:

- $h(\omega)$: Base measure (often $\mathrm{d}\nu = h \mathrm{d}\mu$)
- $\eta(\theta)$: Natural parameters
- $X(\omega)$: Sufficient statistic vector
- $A(\theta)$: Log-normalizer ($A(\theta) = \log \int \mathrm{e}^{\eta(\theta) \cdot X(\omega)} h(\omega) \mathrm{d}\mu(\omega)$).

The exponential representation is not merely algebraic—it induces fundamental geometric properties, that we will be considering throughout this chapter.

For instance, this structure induces in certain well chosen situations a notion of dually flat geometry on Θ. In particular,

- Fisher–Rao metric tensor: $g_{ij}(\theta) = \mathbb{E}_\theta[\partial_i \ell_\theta \partial_j \ell_\theta]$, where $\ell_\theta = \ln \rho_\theta$.
- Dual affine connections ∇ and ∇^* related by g.
- Canonical divergences (Kullback–Leibler, Bregman).

Additionally,

- Geodesics are straight lines in certain well chosen coordinates.
- Pythagorean theorem for divergences.[1]
- Efficient estimation via ∇-autoparallel submanifolds.

A classical example is the normal (Gaussian) family. While the normal distribution indeed constitutes an exponential family, its precise exponential form depends critically on the choice of parametrization.

Exercise 7.4 The Normal/Gaussian distributions on $(\mathbb{R}, \mathbf{B}, \lambda)$, where $\mathbf{B}$ is the Borel algebra on $\mathbb{R}$ and $\lambda[(a, b)] = |b - a|$ the Lebesgue measure is defined by the probability distribution with respect to the Lebesgue measure on $\mathbb{R}$ by:

$$\rho_{\mu,\sigma}(x) = \frac{1}{\sqrt{2\pi}\sigma} \exp -\frac{(x-\mu)^2}{2\sigma^2}, \quad \mu \in \mathbb{R},\ \sigma > 0.$$

Is the normal family exponential? If yes, give a proof.

The specific choice of parametrization affects not only the expression of the density but also the form of the Fisher information metric and the associated affine connections. Thus, determining whether a family of probability distributions is of exponential type—and selecting an appropriate parametrization in cases like the normal family—is fundamental in information geometry. It allows one to exploit powerful geometric techniques to derive insights into the behavior of statistical models, facilitating tasks such as parameter estimation, hypothesis testing, and the study of statistical divergences.

7.7.1 Tangent Space to a Manifold of Probability Distributions

Let $\mathscr{S}$ be a manifold of probability distributions (assumed exponential family) of dimension n. At $P_\theta \in \mathscr{S}$, the tangent space $\mathcal{T}_\theta$ is:

$$\mathcal{T}_\theta = \left\{ u \in L^2(P_\theta) : \mathbb{E}_{P_\theta}[u] = 0,\ u = \sum_{i=1}^{n} u^i \partial_i \ell_\theta \right\}$$

where $\ell_\theta = \ln \rho_\theta$, and $L^2(P_\theta) \equiv L^2(\Omega, \mathbf{S}, P_\theta)$. This space is isomorphic to the span of the score vector $(\partial_i \ell_\theta)_{i=1}^n$.

Remark 7.4 Tangent vectors are equivalence classes of random variables (not signed measures). Since $\rho_\theta > 0$ is μ-almost everywhere, it follows that ℓ_θ is well-defined. Centering $\mathbb{E}_{P_\theta}[u] = 0$ follows from $\partial_i \mathbb{E}_{P_\theta}[1] = 0$.

Exercise 7.5 Prove $\mathbb{E}_{P_\theta}[\partial_i \ell_\theta] = 0$ using:

$$\mathbb{E}_{P_\theta}[\partial_i \ell_\theta] = \int (\partial_i \rho_\theta) \mathrm{d}\mu = \partial_i \underbrace{\int \rho_\theta \mathrm{d}\mu}_{=1} = 0.$$

Exercise 7.6 For Gaussians $\mathcal{N}(\mu, \sigma^2)$, the tangent space $\mathcal{T}_{\mu,\sigma}$ is spanned by:

$$\partial_\mu \ell(x) = \sigma^{-2}(x - \mu)$$
$$\partial_\sigma \ell(x) = \sigma^{-3}(x - \mu)^2 - \sigma^{-1}.$$

Thus $\mathcal{T}_{\mu,\sigma} = \operatorname{span}\{x - \mu, (x - \mu)^2 - \sigma^2\}$. Equivalently:

$$\mathcal{T}_{\mu,\sigma} = \left\{ a\left[(x - \mu)^2 - \sigma^2\right] + b(x - \mu) : a, b \in \mathbb{R} \right\}.$$

7.7.2 The Riemannian Metric

The Fisher–Rao metric at P_θ is defined by the covariance of the score vector:

$$g_{ij}(\theta) = \mathbb{E}_{P_\theta}[\partial_i \ell_\theta \partial_j \ell_\theta] = -\mathbb{E}_{P_\theta}[\partial_i \partial_j \ell_\theta]. \tag{7.6}$$

The dual basis $\{a^i\}_{i=1}^n \subset \mathcal{T}_\theta$ satisfies the orthogonalization condition:

$$\mathbb{E}_{P_\theta}[a^i \partial_j \ell_\theta] = \delta^i_j.$$

The inverse metric tensor (contravariant form) is given by:

$$g^{ij}(\theta) = \mathbb{E}_{P_\theta}[a^i a^j]$$

7.7.3 Parallel Transport in Statistical Manifolds

Consider an infinitesimal perturbation $\theta' = \theta + \mathrm{d}\theta$. Parallel transport is a linear isomorphism:

$$\Pi_\theta^{\theta'} : \mathcal{T}_\theta \to \mathcal{T}_{\theta'}$$

satisfying $\Pi_\theta^\theta = \mathrm{id}$. For $u = u^k \partial_k \in \mathcal{T}_\theta$, its transport to $\mathcal{T}_{\theta'}$ has components:

$$(\Pi_{\theta'}^{\theta} u)^k = u^k + \Gamma_{ij}^k(\theta) u^j \mathrm{d}\theta^i + \mathcal{O}(|\mathrm{d}\theta|^2)$$

where Γ_{ij}^k are Christoffel symbols of the $\nabla^{(1)}$-connection. This defines correspondence between tangent spaces at nearby points.

7.7.4 Covariant Derivative

Let $\pi : E \to \mathcal{S}$ be a vector bundle over the statistical manifold. A covariant derivative is an $\mathbb{R}$-bilinear operator:

$$\nabla : \Gamma(T\mathcal{S}) \times \Gamma(E) \to \Gamma(E), \quad (X, \mathbf{s}) \mapsto \nabla_X \mathbf{s}$$

satisfying the Leibniz rule and where $\mathbf{s}$ is a section of the vector bundle E. For the tangent bundle $E = T\mathcal{S}$ and coordinate basis, the covariant derivative of ∂_j along ∂_i is:

$$\nabla_{\partial_i} \partial_j = \Gamma_{ij}^k(\theta) \partial_k$$

The connection coefficients Γ_{ij}^k measure how basis vectors change under infinitesimal displacement.

7.7.5 Amari–Chentsov Tensor and α-Connections

The Amari–Chentsov tensor is the fully symmetric covariant 3-tensor:

$$\mathbf{T}_{ijk}(\theta) = \mathbb{E}_{P_\theta}[\partial_i \ell_\theta \partial_j \ell_\theta \partial_k \ell_\theta] = \partial_i \partial_j \partial_k \psi(\theta)$$

for exponential families, where ψ is the log-partition function. This tensor generates a family of α-connections through:

$$\Gamma^{(\alpha)k}_{ij} = \Gamma^{(0)k}_{ij} - \frac{\alpha}{2} T_{ij}{}^{k}$$

where $\mathbf{T}_{ij}{}^{k} = g^{kl}\mathbf{T}_{ijl}$ is the (1, 2)-tensor obtained by index raising and

- $$\nabla^{(0)} \quad \text{is the Levi-Civita connection} \quad (\alpha = 0) \tag{7.7}$$
- $$\alpha = 1 \quad \text{yields the } \textbf{exponential connection} \quad \nabla^{(e)} \tag{7.8}$$
- $$\alpha = -1 \quad \text{yields the } \textbf{mixture connection} \quad \nabla^{(m)}. \tag{7.9}$$

Definition 7.12 (Amari–Chentsov tensor) Let $(\mathscr{S}, g)$ be a statistical manifold with Fisher–Rao metric g. The Amari–Chentsov tensor is a fully symmetric covariant 3-tensor $\mathbf{T} \in \Gamma(T^*\mathscr{S}^{\otimes 3})$ defined by:

$$\mathbf{T}(X, Y, Z) = \mathbb{E}_{P_\theta}[(X\ell_\theta)(Y\ell_\theta)(Z\ell_\theta)]$$

where $X, Y, Z \in \Gamma(T\mathscr{S})$ are vector fields, and $X\ell_\theta$ denotes the directional derivative of $\ell_\theta = \ln \rho_\theta$.

The Amari–Chentsov tensor $\mathbf{T}$ is a covariant 3-tensor on a smooth manifold $\mathscr{S}$ which means $\mathbf{T}$ is a multilinear map

$$\mathbf{T} : \Gamma(T\mathscr{S}) \times \Gamma(T\mathscr{S}) \times \Gamma(T\mathscr{S}) \to C^\infty(\mathscr{S}),$$

where $\Gamma(T\mathscr{S})$ denotes the space of smooth vector fields on $\mathscr{S}$.

Properties

- **Full symmetry**:
$$\mathbf{T}(X, Y, Z) = \mathbf{T}(Y, X, Z) = \cdots$$
for all $X, Y, Z \in \Gamma(T\mathscr{S})$.
- C^∞**-linearity in each argument**: For all smooth functions $f \in C^\infty(\mathscr{S})$,
$$\mathbf{T}(fX, Y, Z) = f \cdot \mathbf{T}(X, Y, Z).$$

Definition 7.13 (Musical isomorphism) The metric g induces a bundle isomorphism $\sharp : T^*\mathscr{S} \to T\mathscr{S}$ defined by:

$$g(\omega^\sharp, Z) = \omega(Z) \quad \forall Z \in \Gamma(T\mathscr{S})$$

for $\omega \in \Gamma(T^*\mathscr{S})$.

Application to $\mathbf{T}(X, Y, \cdot)$

Fix two vector fields $X, Y \in \Gamma(T\mathscr{S})$. We define the 1-form $\omega_{XY} \in \Gamma(T^*\mathscr{S})$.[2]

The Amari–Chentsov tensor is given by $T \in \Gamma(T^*\mathscr{S}^{\otimes 3})$. The map

$$\omega_{XY} : \Gamma(T\mathscr{S}) \to C^\infty(\mathscr{S}), \quad \omega_{XY}(Z) := \mathbf{T}(X, Y, Z)$$

satisfies the linearity property and is $C^\infty(\mathscr{S})$-linear:

$$\omega_{XY}(fZ) = f \cdot \omega_{XY}(Z), \quad \forall f \in C^\infty(\mathscr{S}).$$

Thus ω_{XY} is a smooth 1-form, i.e., $\omega_{XY} \in \Gamma(T^*\mathscr{S})$.

In coordinates $\{\theta^i\}$ we have:

$$\omega_{XY}(Z) = \mathbf{T}_{ijk} X^i Y^j Z^k,$$

which is linear in Z^k and with coefficients $c_k = \mathbf{T}_{ijk} X^i Y^j$, confirming that $\omega_{XY} = \left(\mathbf{T}_{ijk} X^i Y^j\right) \mathrm{d}\theta^k$.

The sharp operator yields a vector field $A_{XY} = \omega_{XY}^\sharp$ satisfying:

$$g(A_{XY}, Z) = \mathbf{T}(X, Y, Z), \quad \forall Z \in \Gamma(T\mathscr{S}).$$

Coordinate Representation

In local coordinates $\{\theta^i\}$ on $\mathscr{S}$:

1. **Tensor Components**: $\mathbf{T}_{ijk} = \mathbf{T}(\partial_i, \partial_j, \partial_k)$
2. **Vector Field Decomposition**: $X = X^i \partial_i, Y = Y^j \partial_j$
3. **1-Form Components**: $\omega_{XY} = \mathbf{T}_{ijk} X^i Y^j \mathrm{d}\theta^k$
4. **Sharp Operator Action**:
$$A_{XY}^k = g^{kl}(\mathbf{T}_{ijl} X^i Y^j)$$
where g^{kl} is the inverse metric tensor.

We provide a basis-free proof. Let $A_{XY} = \omega_{XY}^\sharp$. For any Z:

$$g(A_{XY}, Z) = \omega_{XY}(Z) = \mathbf{T}(X, Y, Z)$$

[2] We can think of $\mathbf{T}(X, Y, \cdot)$ as the "freezing" of a trilinear form to a linear form.

In coordinates:

$$g_{km} A^m_{XY} Z^k = \mathbf{T}_{ijk} X^i Y^j Z^k$$

Equating coefficients for all Z:

$$g_{km} A^m_{XY} = \mathbf{T}_{ijk} X^i Y^j$$

Applying g^{lk}:

$$\delta^l_m A^m_{XY} = g^{lk} \mathbf{T}_{ijk} X^i Y^j \implies A^l_{XY} = g^{lk} \mathbf{T}_{ijk} X^i Y^j$$

confirming $A^k_{XY} = g^{kl} \mathbf{T}_{ijl} X^i Y^j$.

Role

The difference between the α- and $-\alpha$-connections is proportional to this vector field:

$$\nabla^{(\alpha)}_X Y - \nabla^{(-\alpha)}_X Y = \alpha \cdot \underbrace{\mathbf{T}(X, Y, \cdot)^\sharp}_{\text{vector field}}.$$

This measures the asymmetry induced by the Amari–Chentsov tensor.

The covariant exterior derivative d_∇ extends the exterior derivative to tensor-valued forms. For a covariant 3-tensor $\mathbf{T}$:

$$(d_\nabla \mathbf{T})(X, Y, Z, W) = (\nabla_X \mathbf{T})(Y, Z, W) - (\nabla_Y \mathbf{T})(X, Z, W).$$

The condition $d_\nabla \mathbf{T} = 0$ means that $\mathbf{T}$ is covariantly closed.

Consequences

- For statistical manifolds, $d_\nabla \mathbf{T} = 0$ implies $\mathbf{T}$ is the third derivative of a potential, e.g.:

$$\mathbf{T}_{ijk} = \partial_i \partial_j \partial_k \psi.$$

- It ensures that the α and $-\alpha$-connections are dual with respect to g.

Why These Objects Appear in Information Geometry

Concept	Role in statistical manifolds
$\mathbf{T}(X, Y, \cdot)^\sharp$	Quantifies how the α-connections deviate from the Levi–Civita connection via statistical skewness
$d_\nabla \mathbf{T} = 0$	Ensures $\mathbf{T}$ is geometrically integrable (arises from a potential ψ in exponential families)

To summarize the α-connections satisfy:

$$\nabla_X^{(\alpha)} Y - \nabla_X^{(-\alpha)} Y = \alpha \cdot \mathbf{T}(X, Y, \cdot)^\sharp$$
$$d_\nabla \mathbf{T} = 0 \quad \text{(Codazzi condition)},$$

and quantify information curvature beyond Riemannian geometry.

7.7.6 Statistical Inference and α-Connections

The goal of this subsection is to demonstrate how the differential-geometric tools introduced above find concrete applications in statistics. We do not attempt to review classical statistical concepts, which we assume to be familiar to the reader:

Core Components of Statistical Inference

Probability Model Consider a family of probability distributions

$$\mathscr{S} = \{P_\theta : \theta \in \Theta\},$$

parameterized by $\theta \in \Theta$.

Observed Data Let

$$x = (x_1, \ldots, x_n)$$

be an i.i.d. sample drawn from the true (unknown) distribution P_{θ_0}.

Inferential Goals Statistical inference typically involves three main tasks:

- **Estimation**: Approximate the true parameter θ_0 using an estimator $\hat{\theta}(x)$ (e.g., the maximum likelihood estimator).
- **Hypothesis Testing**: Decide between the competing hypotheses

$$H_0 : \theta \in \Theta_0 \qquad \text{versus} \qquad H_1 : \theta \in \Theta_1.$$

- **Uncertainty Quantification**: Construct confidence regions for θ_0.

△ Statistical inference is the mathematical framework for extracting knowledge from data by estimating parameters, testing hypotheses, and quantifying uncertainty in probabilistic models. Its profound connection to differential geometry is revealed through α-connections, which quantify how information accumulates along different paths in the space of probability distributions.

7.7.6.1 Geometry of Inference

In this construction we consider as above the statistical model $\mathscr{S} = \{P_\theta : \theta \in \Theta\}$ and some observed data which is a sequence of i.i.d. random variables $\mathbf{x} = (x_1, \ldots, x_n)$ from an unknown distribution $P_\theta \in \mathscr{S}$.

The sequence:

$$P_{\hat{\theta}_1} \to P_{\hat{\theta}_2} \to \cdots \to P_{\hat{\theta}_N}$$

represents the evolution of estimated probability distributions as we incrementally add data points.

What Is Meant by This: Step-By-Step Construction

1. **Start with One Data Point**: Compute the estimate $\hat{\theta}_1$ using only x_1. This defines a distribution $P_{\hat{\theta}_1} \in \mathscr{S}$.

$$\text{Example (Gaussian model):} \qquad P_{\hat{\theta}_1} = \mathcal{N}(x_1, \sigma_0^2).$$

2. **Add a Second Data Point**: Compute $\hat{\theta}_2$ using $\{x_1, x_2\}$ and move to the new distribution $P_{\hat{\theta}_2}$.

$$P_{\hat{\theta}_2} = \mathcal{N}\Big(\frac{x_1 + x_2}{2}, \sigma_1^2\Big).$$

3. **Continue to N Points**: For each n, update $\hat{\theta}_n$ using $\{x_1, \ldots, x_n\}$, obtaining the distribution $P_{\hat{\theta}_n}$ at step n.
4. **Result**: This procedure generates a discrete path on the statistical manifold:

$$P_{\hat{\theta}_1} \longrightarrow P_{\hat{\theta}_2} \longrightarrow \cdots \longrightarrow P_{\hat{\theta}_N},$$

where each arrow "$\to$" represents an update via the estimator $\hat{\theta}_n$.

When we estimate θ from data, we traverse paths in $\mathscr{S}$.

1. Using the **Fisher–Rao Metric**
The *Fisher–Rao metric* $g_{ij}(\theta)$ (see Eq. 7.6) measures the "distance" between nearby probability distributions. It governs the lower bound on the variance of unbiased estimators through the so-called Cramér–Rao inequality:

$$\text{Var}(\hat{\theta}) \geq g^{-1}(\theta),$$

where $g^{-1}(\theta)$ denotes the inverse of the Fisher information matrix.

2. **α-Connections**
   The family of *α-connections* $\nabla^{(\alpha)}$ determines the “straight paths” (geodesics) for estimation on the statistical manifold. The Christoffel symbols $\Gamma_{ij}^{(\alpha)k}$ associated with these connections are given by

$$\Gamma_{ij}^{(\alpha)k} = \Gamma_{ij}^{(0)k} - \frac{\alpha}{2} g^{kl} T_{ijl},$$

   where $\Gamma_{ij}^{(0)k}$ is the Levi–Civita connection of g_{ij} and T_{ijl} is the Amari–Chentsov cubic tensor. For the table below see Eqs. 7.7–7.9.

Connection	α-value	Statistical meaning
$\nabla^{(e)}$	1	Maximum likelihood estimation, exponential families, asymptotically efficient
$\nabla^{(m)}$	-1	Robust estimation, mixture models, outlier-resistant
$\nabla^{(0)}$	0	Bayesian posterior mode estimation, balances bias–variance trade-off

In information geometry, the coefficients of the α-connection in covariant form are defined as:

$$\Gamma_{ijk}^{(\alpha)}(\theta) = \mathbb{E}_\theta\left[\left(\partial_i\partial_j\ell(x;\theta) + \frac{1-\alpha}{2}\partial_i\ell(x;\theta)\partial_j\ell(x;\theta)\right)\partial_k\ell(x;\theta)\right]$$

where $\ell(x;\theta) = \ln p(x;\theta)$ is the log-likelihood function.

7.7.6.2 Summary

1. **Special Cases**:

   - $\alpha = 1$: Exponential connection $\Gamma_{ijk}^{(1)} = \mathbb{E}_\theta[\partial_i\partial_j\ell \cdot \partial_k\ell]$
   - $\alpha = -1$: Mixture connection $\Gamma_{ijk}^{(-1)} = \Gamma_{ijk}^{(1)} + T_{ijk}$
   - $\alpha = 0$: Levi–Civita connection $\Gamma_{ijk}^{(0)} = \Gamma_{ijk}^{(1)} + \frac{1}{2}T_{ijk}$.

2. **Statistical Meaning**:

$$\Gamma_{ijk}^{(1)} = \text{Covariance}(\partial_i\partial_j\ell, \partial_k\ell)$$
$$T_{ijk} = \text{Skewness}(\partial_i\ell, \partial_j\ell, \partial_k\ell)$$

3. **Duality**: $\Gamma_{ijk}^{(\alpha)} + \Gamma_{ijk}^{(-\alpha)} = 2\Gamma_{ijk}^{(0)}$
   It can be proved that the α-connection quantifies how estimation efficiency varies with the underlying geometry (see works of Amari–Chentsov for details):

- The exponential connection ($\alpha = 1$) governs:
  - Maximum likelihood estimation efficiency
  - Cramér–Rao bound attainment
  - Exponential family flatness.
- The mixture connection ($\alpha = -1$) characterizes:
  - Robust estimation
  - mixture model geometry
  - Projection theorems.
- General α interpolates between frequentist ($\alpha = 1$) and Bayesian ($\alpha = -1$) paradigms.

The interplay between these structures underpins modern statistical inference and highlights the deep geometric nature of estimation, testing, and uncertainty quantification.

7.8 Hints and Solutions to Exercises

Exercise 7.1. It is easy to show it using basic set theory.

Exercise 7.2.

1. **Second-Countability of X:**
   By definition, X is second-countable, meaning it has a countable basis $\mathcal{B} = \{B_n\}_{n\in\mathbb{N}}$ for its topology. That is, every open set in X can be written as a union of elements of $\mathcal{B}$.
2. **Open Cover $\mathcal{U}$:**
   Let $\mathcal{U} = \{U_\alpha\}_{\alpha\in I}$ be an open cover of X. This means:
   $$X = \bigcup_{\alpha\in I} U_\alpha.$$
3. **Constructing a Countable Subcover:**
   For each $x \in X$, there exists some $U_\alpha \in \mathcal{U}$ such that $x \in U_\alpha$.
   Since $\mathcal{B}$ is a basis, there exists some $B_{n_x} \in \mathcal{B}$ such that:
   $$x \in B_{n_x} \subseteq U_\alpha.$$
   Let $\mathcal{B}' = \{B_{n_x} \mid x \in X\}$. This is a subset of $\mathcal{B}$, and since $\mathcal{B}$ is countable, $\mathcal{B}'$ is also countable.

4. **Extracting a Countable Subcover**:
   For each $B_{n_x} \in \mathcal{B}'$, choose one U_α such that $B_{n_x} \subseteq U_\alpha$. Let $\mathcal{U}'$ be the collection of all such U_α.
   Since $\mathcal{B}'$ is countable, $\mathcal{U}'$ is also countable.
5. **Verification that $\mathcal{U}'$ Is a Cover**:
   For any $x \in X$, there exists $B_{n_x} \in \mathcal{B}'$ such that $x \in B_{n_x} \subseteq U_\alpha$ for some $U_\alpha \in \mathcal{U}'$. Thus, $x \in U_\alpha$, and since x was arbitrary, $\mathcal{U}'$ covers X.
6. **Conclusion**:
   We have constructed a countable subcover $\mathcal{U}'$ of $\mathcal{U}$, proving the theorem.

Exercise 7.3.

If $\inf_{x\in A, y\in B} d(x, y) = \delta > 0$ then the function f can be chosen to be uniformly continuous.

Let $f(x) = d(x, A)/[d(x, A) + d(x, B)]$. It is easy to verify that f satisfies both conditions 1 and 2. The other part of the theorem follows from the fact that $d(a, X) + d(x, B) \geq \delta$ and from the following fact that the function $d(x, A)$ satisfies the inequality

$$|d(x, A) - d(y, A)| \leq d(x, y).$$

In particularly, $d(x, A)$ is uniformally continuous.

Exercise 7.8. The proof is omitted since it is considered in Exercise 7.2.

Exercise 7.5. The proof is omitted since it is a straightforward calculation.

Exercises 7.4 and 7.6. By an appropriate change of variables, for the Gaussian, this density can be rewritten in the exponential family form, revealing its natural parameters and sufficient statistics.

Categorical Structures in Probability and Measure Theory 8

In this chapter, we embark on a journey into probability and statistics in a manner that may initially surprise the reader: through the lens of *philosophy*. Rather than taking the conventional route of axiomatic probability theory, we begin with a deeper reflection on the nature of mathematical structures and their underlying symmetries. Our point of departure is a generalization of Felix Klein's revolutionary insight–his *Erlangen Program*–which reinterpreted geometry as the study of figures and their congruences under transformation groups.

We invite the reader to consider an analogous perspective in the realm of probability distributions. Just as Klein's figures reside in a structured mathematical world shaped by transformations, we view probability distributions as abstract entities within a broader conceptual space. This space, much like Plato's world of ideas, consists of idealized probabilistic objects related by structural equivalences. These equivalences, or congruences, form the foundation upon which we build a categorical framework for probability theory.

This shift in perspective naturally leads us to the introduction of categories of probability distributions, where morphisms arise from Markov kernels–stochastic mappings that embody the very notion of congruence in this setting. In this way, Markov kernels serve as the probabilistic analogues of Klein's geometric transformations, revealing deeper structures in statistical theory that might otherwise remain hidden.

Through this approach, we illuminate the rich interplay between probability, statistics, and abstract mathematical structures, setting the stage for a unifying perspective that extends beyond classical interpretations.

© The Author(s), under exclusive license to Springer Nature Switzerland AG 2026

N. C. Combe et al., *Exploring Information Geometry*, Frontiers in Mathematics,
https://doi.org/10.1007/978-3-032-23054-6_8

8.1 Philosophical Approach

8.1.1 What Is the Aim of Geometry?

Let us begin by considering a broad and fundamental question:

what is, in general, the aim of geometry?

One can state succinctly that geometry is primarily concerned with:

- The study of spaces.
- The transformations of these spaces, namely, their motions or the actions of groups.
- Entire classes of geometric objects.

The foundations of geometry, as formulated by Klein, are based on the study of geometric figures and the transformations that act upon them. This perspective, known as *Erlangen's program*, emphasizes the structural relationships between figures rather than their specific realizations. In its classical setting, this formalism provides a natural framework for understanding symmetries, invariants, and classification problems within geometry.

For us, however, we will think of those (Klein) figures as figures *in the sense of Plato*. Plato's world of geometric figures refers to his philosophical concept of the *World of Forms*, where perfect and unchanging geometric shapes exist, independently of the physical world.

In Plato's philosophy, as described in dialogues such as the *Republic* and the *Phaedo*, mathematical objects: circles, triangles, and other ideal forms do not exist merely as physical approximations but as pure, abstract entities in a higher, non-material realm.[1]

It is precisely *this concept* of Plato's world of figures/forms which is used in Klein's perception and which we will use also to describe geometric structures in probability and measure theory.

In the present work, we extend this viewpoint beyond its original domain by considering a formalism in which Klein's approach is adapted to the realm of probability theory and statistics. The motivation for such a generalization arises naturally: if one seeks to construct a statistical geometry, it is necessary to first identify the appropriate notion of *geometric figures* and the corresponding transformations that preserve the relevant structures.

Our approach follows a twofold strategy. First, we identify probabilistic objects that play the role of geometric figures in this generalized setting. These may include, for instance, families of probability distributions, families of distributions, or spaces of statistical models.

[1] In *The Republic,* Plato's "Allegory of the Cave" illustrates how ordinary humans perceive mere shadows (imperfect physical objects), while true philosophers seek to understand the Forms, including perfect geometric structures. A drawn triangle or sphere in the physical world is an imperfect shadow of the true ideal triangle or sphere that exists in the intelligible realm.

Second, we introduce transformations between these objects that preserve key statistical or probabilistic properties.

The fundamental problem then becomes:

how does one define and classify these transformations, and what structures do they induce?

> From this perspective, Markov categories and other categories (introduced in this chapter and the next chapter) emerge as natural generalizations of geometric figures in Klein's sense, and their transforms. These transformations (or actions)–given by statistical morphisms, stochastic maps, or Markov kernels–serve as the analogues of geometric transformations in classical geometry. This interpretation provides a unifying language in which statistical inference, decision theory, and information geometry can be understood in terms of an overarching geometric framework.

The goal of this work is to explore this formalism systematically and to establish the necessary mathematical structures that allow for a coherent treatment of statistical geometry. In particular, we shall investigate the role of categorical constructions, invariants under transformations, and the emergence of new geometric notions adapted to probabilistic contexts. Ultimately, our approach seeks to trivialize Klein's original vision in such a way that its extension to statistics appears as a natural and inevitable development.

The Klein figure formalism admits a natural extension to probabilistic contexts, particularly in the study of parametrized families of probability distributions. This adaptation establishes a coherent theoretical framework that facilitates systematic analysis and operational flexibility within such probabilistic structures.

8.1.2 Towards a Generalization of Klein's Plato-Figures

The most elementary class of geometric objects consists of subsets of a given space, referred to by Felix Klein as *figures*.

- Two figures A and B are said to be *congruent* if there exists a motion mapping A onto B, along with an inverse motion mapping B onto A. The geometer's principal concern is to investigate the properties of figures that remain invariant under such motions–those properties that are identical for all congruent figures.
- The central problem thus becomes the determination of *invariants* of figures, namely, real-valued functions that take equal values on congruent figures. A complete system of such invariants allows for a full **classification of geometric configurations.**

Since any figure can be regarded as a set of points, one is naturally led to a **generalization of this concept** of geometry–extending beyond the study of figures in the classical sense. Alongside figures in Klein's sense, which are simply sets, one may consider *parametrized sets*, thereby introducing a new geometric framework to examine their intrinsic properties.

Let Θ be an index set parametrizing such a collection, and let us construct an epimorphism from this parameter set onto the space under consideration. That is, for each parameter value $\theta \in \Theta$, a corresponding point of the considered figure is assigned. In general, a single point of the figure may be associated with multiple parameter values, reflecting a richer underlying structure.

Following Klein's approach to geometry, we may consider parametrized sets as Klein figures, where the parametrization is defined in a specific way. Such examples include:

- finite ordered sequences of points;
- parametrized smooth curves or surfaces;
- smooth manifolds parametrized by locally smooth coordinates–these can be treated both as points or as smooth surfaces.

Furthermore, in the scope of generalizing objects and structures, we can recall what a category is, as we will need it later.

8.1.3 Categories as a Classifying Tool

8.1.3.1 Categories

A category is a mathematical structure that captures relationships between objects in an abstract way. At its core, a category consists of objects and morphisms (arrows) between them, which satisfy rules of composition and identity. One should think of a category not just as a formalism but a lens through which mathematics reveals its hidden symmetries and harmonies. We previously introduced this concept in Sect. 6.3.2 while familiarizing the reader with alternative, more algebraic tools.

Formally, a category $\mathscr{C}$ consists of:

- A collection of objects, denoted $\mathrm{Ob}(\mathscr{C})$.
- For each pair of objects $X, Y \in \mathrm{Ob}(\mathscr{C})$, a set $\mathrm{Hom}_{\mathscr{C}}(X, Y)$ of morphisms $f : X \to Y$.
- For every triple of objects X, Y, Z, a **composition map**

$$\circ : \mathrm{Hom}_{\mathscr{C}}(Y, Z) \times \mathrm{Hom}_{\mathscr{C}}(X, Y) \to \mathrm{Hom}_{\mathscr{C}}(X, Z),$$

  written as $(g, f) \mapsto g \circ f$.
- For each object X, an **identity morphism** $\mathrm{id}_X \in \mathrm{Hom}_{\mathscr{C}}(X, X)$.

These satisfy the following axioms:
(1) Associativity: For all $f : X \to Y, g : Y \to Z, h : Z \to W$,

$$h \circ (g \circ f) = (h \circ g) \circ f.$$

(2) Identity: For every morphism $f : X \to Y$,

$$f \circ \mathrm{id}_X = f \quad \text{and} \quad \mathrm{id}_Y \circ f = f.$$

8.1.3.2 Covariant Functor

A **covariant functor** $F : \mathcal{C} \to \mathcal{D}$ consists of:

- **An Object Assignment:** For each object $X \in \mathcal{C}$, an object $F(X) \in \mathcal{D}$.
- **A Morphism Assignment:** For each morphism $f : X \to Y$ in $\mathcal{C}$, a morphism

$$F(f) : F(X) \to F(Y)$$

in $\mathcal{D}$.

These assignments satisfy:

- **Preservation of Identity:**

$$F(\mathrm{id}_X) = \mathrm{id}_{F(X)} \quad \text{for all } X \in \mathcal{C}.$$

- **Preservation of Composition:** For morphisms $f : X \to Y$ and $g : Y \to Z$ in $\mathcal{C}$:

$$F(g \circ f) = F(g) \circ F(f).$$

Diagram for Covariant Functor

$$\begin{array}{ccc} X & \xrightarrow{f} & Y \\ \downarrow & & \downarrow \\ F(X) & \xrightarrow{F(f)} & F(Y) \end{array}$$

Example: Power Set Functor

The power set functor $P : \mathbf{Set} \to \mathbf{Set}$:

- On objects: $S \mapsto P(S)$ (the power set of S).
- On morphisms: $f : S \to T$ maps to

$$P(f) : P(S) \to P(T), \qquad P(f)(A) = f(A) \quad \text{for } A \subseteq S.$$

8.1.3.3 Contravariant Functor

A **contravariant functor** $F : \mathcal{C} \to \mathcal{D}$ consists of:

- **An Object Assignment:** Same as a covariant functor.
- **A Morphism Assignment:** For each morphism $f : X \to Y$ in $\mathcal{C}$, a morphism

$$F(f) : F(Y) \to F(X)$$

  in $\mathcal{D}$ (note: the direction is reversed).

  These assignments satisfy:

- **Preservation of Identity:**

$$F(\mathrm{id}_X) = \mathrm{id}_{F(X)}.$$

- **Reversal of Composition:** For morphisms $f : X \to Y$ and $g : Y \to Z$:

$$F(g \circ f) = F(f) \circ F(g).$$

Diagram for Contravariant Functor

$$\begin{array}{ccc} X & \xrightarrow{f} & Y \\ \downarrow & & \downarrow \\ F(X) & \xleftarrow{F(f)} & F(Y) \end{array}$$

8.1.3.4 Example: Dual Space Functor

The dual space functor $(-)^* : \mathbf{Vect}_k \to \mathbf{Vect}_k$:

- On objects: $V \mapsto V^*$ (the dual space of V).
- On morphisms: a linear map $f : V \to W$ maps to

$$f^* : W^* \to V^*, \qquad f^*(\varphi) = \varphi \circ f.$$

8.1.3.5 Unified Perspective

A contravariant functor $F : \mathcal{C} \to \mathcal{D}$ is equivalent to a covariant functor

$$F : \mathcal{C}^{op} \to \mathcal{D},$$

where $\mathcal{C}^{op}$ is the opposite category (all morphisms reversed).

Summary

Property	Covariant functor	Contravariant functor
Direction	Preserves: $X \to Y \mapsto F(X) \to F(Y)$	Reverses: $X \to Y \mapsto F(Y) \to F(X)$
Composition	$F(g \circ f) = F(g) \circ F(f)$	$F(g \circ f) = F(f) \circ F(g)$

8.2 Generalization of Klein's (Plato)-Figures/Congruences

The work of N.N. Chentsov and E.A. Morozova on Klein figures in Markov geometry is a deep topic in geometric probability and information geometry. Their framework connects Markov dynamics, statistical manifolds, and geometric structures inspired by Felix Klein's Erlangen Program. Below is a structured explanation:

Chentsov and Morozova extended Klein's idea to Markov geometry, where the "space" is a statistical manifold (a space of probability distributions), and the "transformations" are Markov morphisms (stochastic maps).

8.2.1 Generalized Klein Figures

Klein's notion of geometric figures is extended to probability theory through the following examples of *generalized Klein figures*:

- spaces of probability distributions of a fixed type (e.g., Gaussian distributions on $\mathbb{R}^n$);
- dominated statistical models, i.e., particularly finite families of probability distributions defined on finite σ-algebras.

(Any finite or countable family $\{P_k\}$ of probability measures on a finite σ-algebra is dominated, as it admits a dominating measure such as $\sum_k c_k P_k$ where $c_k > 0$ and $\sum_k c_k = 1$.)

8.2.2 Markov Transformations

A statistical manifold is a space of probability distributions (e.g., Gaussian distributions, multinomial distributions). A Markov morphism is a stochastic map (e.g., coarse-graining, noise addition) that transforms one distribution into another.

Just as Euclidean geometry studies invariants under rigid motions, Markov geometry studies invariants under Markov morphisms.

Chentsov (1972) proved that, up to scaling, the Fisher information metric is the unique Riemannian metric on a statistical manifold that is invariant under Markov morphisms. This makes the Fisher metric the natural "distance" for statistical inference in this framework.

8.2.3 Generalized Klein Figures and Transformations

Chentsov and Morozova introduced Klein figures as geometric objects that encode invariant properties under Markov morphisms.

A Klein figure is a geometric structure (e.g., a simplex, cone, or polytope) equipped with a Markov-invariant structure. Examples include the probability simplex (space of discrete distributions) with the Fisher metric. Another important object is the cone of positive definite matrices (Gaussian case) with the affine-invariant metric. These figures are "universal" in the sense that they appear naturally under Markov morphisms.

Klein figures serve as building blocks for constructing more complex statistical manifolds. They generalize the idea of symmetric spaces (like spheres, hyperbolic spaces) to the stochastic setting.

Using the Klein figure formalism, an essential ingredient in constructing a geometric framework for probabilistic or statistical quantities is the study of transformations acting on generalized Klein figures. We now introduce such transformations.

Definition 8.1 Let $(\Omega, \mathbf{S})$ and $(\Omega', \mathbf{S}')$ be measurable spaces. A *Markov transition kernel* (also called a *Markov morphism*) is a mapping

$$\Pi : \Omega \times \mathbf{S}' \to [0, 1]$$

satisfying:

1. For each $\omega \in \Omega$, $\Pi(\omega, \cdot)$ is a probability measure on $(\Omega', \mathbf{S}')$.
2. For each $A \in \mathbf{S}'$, $\Pi(\cdot, A)$ is $\mathbf{S}$-measurable.

Every such kernel induces a transformation on generalized Klein figures:

$$T_\Pi : \mathscr{P}(\Omega, \mathbf{S}) \to \mathscr{P}(\Omega', \mathbf{S}')$$

defined by

$$T_\Pi(P)(A) = \int_\Omega \Pi(\omega, A) dP(\omega), \quad \forall A \in \mathbf{S}',$$

where $\mathscr{P}(\cdot)$ denotes the space of probability measures. This T_Π maps a probability measure P to its *pushforward* under the Markov kernel.

This kernel describes a Markovian transition from $(\Omega, \mathbf{S})$ to $(\Omega', \mathbf{S}')$. We now illustrate the concept of congruences (mentioned in Sect. 8.1.2) through a fundamental statistical example.

Example 8.1 (*Parameter estimation for normal distributions*) Consider estimating parameters μ (mean) and σ^2 (variance) of a normal distribution based on N independent observations $\{x^{(1)}, \ldots, x^{(N)}\}$. The density is:

$$p(x; \mu, \sigma) = \frac{1}{\sqrt{2\pi}\,\sigma} \exp\left(-\frac{(x-\mu)^2}{2\sigma^2}\right), \quad \sigma > 0.$$

Geometrically, the space of all probability measures absolutely continuous with respect to Lebesgue measure on $\mathbb{R}$ is infinite-dimensional. The family of normal distributions $\{p(\cdot; \mu, \sigma) \mid \mu \in \mathbb{R}, \sigma > 0\}$ forms a 2-dimensional submanifold (a statistical manifold) embedded in this ambient space. This manifold is diffeomorphic to the half-plane $\mathbb{R} \times (0, \infty)$.

Theorem 8.1 (Chentsov's Uniqueness Theorem (1972)) *Let $\mathcal{P}(\Omega)$ be the space of probability distributions on a finite sample space Ω with $n \geq 3$ outcomes, equipped with a Riemannian metric g. If g satisfies:*

1. ***Invariance Under Markov Morphisms:*** *For every Markov embedding $T \colon \mathcal{P}(\Omega) \hookrightarrow \mathcal{P}(\Omega')$, the pullback metric satisfies $T^*g = g$.*
2. ***Consistency with Coarse-Graining:*** *The metric restricts properly to sub-simplices corresponding to marginal distributions.*

Then g must be a constant multiple of the Fisher information metric:

$$g_{Fisher}(\theta) = \alpha \sum_{i=1}^{n-1} \frac{d\theta_i \otimes d\theta_i}{\theta_i}, \quad \alpha > 0,$$

where $\theta = (\theta_1, ..., \theta_{n-1})$ are the canonical coordinates on the probability simplex Δ^{n-1}.

Key Ideas of Proof 1. **Markov embeddings** generate all possible conditional probability maps between simplices of different dimensions.

2. The Fisher metric is the only metric that scales correctly under **coarse-graining operations** (merging outcomes) and **splitting operations** (refining outcomes).

3. The proof uses the **representation theory** of stochastic maps acting on the tangent spaces of the probability simplex. □

8.2.4 Klein's Congruences and the Universal Property

The following theorem characterizes when two families of probability distributions are **congruent**, meaning they share a common parameterization through a finite auxiliary space $\mathcal{E}$. This structural equivalence resembles universal properties in category theory.

Theorem 8.2 (Universal property) *Consider two parametric families of probability measures on measurable spaces $(\Omega^{(i)}, \mathbf{S}^{(i)})$ for $i = 1, 2$ defined as $\mathcal{F}_i = \{P_\theta^{(i)} : \theta \in \Theta\}$.*

*These families are **congruent** if there exist:*

1. *a finite measurable space $(\mathcal{E}, \mathcal{B})$, where $\mathcal{B}$ is the power set σ-algebra*[2]*;*
2. *measurable quotient maps $\varphi_i : \Omega^{(i)} \to \mathcal{E}$ that are surjective;*
3. *a common parametric family $\{R_\theta : \theta \in \Theta\}$ of probability measures on $(\mathcal{E}, \mathcal{B})$;*
4. *Markov transition kernels $Q^{(i)} : \mathcal{E} \times \mathbf{S}^{(i)} \to [0, 1]$ satisfying*

$$P_\theta^{(i)}(A) = \int\limits_{\mathcal{E}} Q^{(i)}(\epsilon, A) R_\theta(d\epsilon) \quad \forall A \in \mathbf{S}^{(i)}, \theta \in \Theta$$

$$Q^{(i)}(\epsilon, \varphi_i^{-1}(\{\epsilon\})) = 1 \quad \forall \epsilon \in \mathcal{E}.$$

This implies the following commutative diagram of measurable spaces:

$$(\Omega^{(1)}, \mathbf{S}^{(1)}) \xrightarrow{\varphi_1} (\mathcal{E}, \mathcal{B}) \xrightarrow{\varphi_2^{-1}} (\Omega^{(2)}, \mathbf{S}^{(2)}), \qquad (\Omega^{(1)}, \mathbf{S}^{(1)}) \xrightarrow{\varphi_2 \circ \psi} (\Omega^{(2)}, \mathbf{S}^{(2)})$$

where $\psi : \Omega^{(1)} \to \Omega^{(2)}$ is a measurable isomorphism satisfying $\varphi_2 \circ \psi = \varphi_1$.[34]

[2] Since $\mathcal{E}$ is finite (as explicitly stated in the theorem), its power set is a valid σ-algebra (closed under complementation, countable unions/intersections) and the finest possible σ-algebra (all subsets are measurable).

[3] The critical condition $Q^{(i)}(\epsilon, \varphi_i^{-1}(\epsilon) = 1$ requires that singletons ϵ are measurable (so their preimages $\varphi_i^{-1}(\epsilon)$ are measurable in $(\Omega^{(i)}, \mathbf{S}^{(i)})$). The power set σ-algebra guarantees this.

Example 8.2 If $\mathscr{E} = \{a, b, c\}$, then the σ-algebra $\mathscr{B}$ on $\mathscr{E}$ is

$$\mathscr{B} = \{\emptyset, \{a\}, \{b\}, \{c\}, \{a, b\}, \{a, c\}, \{b, c\}, \{a, b, c\}\}.$$

We discuss key implications in the theorem.

- **Preimages Are Measurable:** for every $\epsilon \in \mathscr{E}$,

$$\varphi_i^{-1}(\{\epsilon\}) \in \mathbf{S}^{(i)}.$$

- **Kernels Are Well-Defined:** the transition kernel $Q^{(i)}(\epsilon, \cdot)$ assigns probabilities to arbitrary subsets of $\mathscr{E}$.
- **No Measurability Obstructions:** the finite structure endowed with the maximal σ-algebra avoids technical complications.

Example 8.3 (*Congruence of probability simplices*) We illustrate Theorem 8.2 through a geometric construction involving congruent families of probability distributions represented within simplices. Consider two probability simplices:

- The first simplex $\mathbf{\Sigma}_1 := \Delta(\Omega)$ corresponds to distributions on a sample space $\Omega = \{\omega_1, \ldots, \omega_n\}$, with vertices $\mathbf{e}_1, \ldots, \mathbf{e}_n$ representing Dirac measures at each outcome.
- The outcomes are partitioned into m disjoint subsets $E_1, \ldots, E_m$ where $\bigcup_{k=1}^m E_k = \Omega$.

For each subset E_k, consider the face $\boldsymbol{F}_k \subseteq \mathbf{\Sigma}_1$ spanned by vertices $\{\mathbf{e}_i : \omega_i \in E_k\}$. Within $\boldsymbol{F}_k$, we select a point P_k with barycentric coordinates corresponding to a probability distribution supported on E_k. These distributions are chosen such that:

- $P_k(\omega_i) > 0$ for all $\omega_i \in E_k$;
- $P_k(\omega_j) = 0$ for $\omega_j \notin E_k$;
- All but one P_k are interior points of their respective faces.

The convex hull of $\{P_1, \ldots, P_m\}$ forms an $(m-1)$-dimensional subsimplex $\mathbf{\Sigma}_1' \subseteq \mathbf{\Sigma}_1$. This construction establishes a congruence with a second simplex $\mathbf{\Sigma}_2$ via the finite quotient space $\mathscr{E} = \{1, \ldots, m\}$ where:

- the measurable mapping $\varphi_i : \Omega^{(i)} \to \mathscr{E}$ sends $\omega \mapsto k$ if $\omega \in E_k$;
- the common family R_θ is parameterized by barycentric coordinates in $\mathscr{E}$;
- transition distributions $Q^{(i)}$ recover the P_k distributions.

[4] For the quotient maps $\varphi_i : \Omega^{(i)} \to \mathscr{E}$ to be measurable, every subset of $\mathscr{E}$ (including fibers of points) must be measurable. Only the power set ensures this universally.

The commutative diagram:

$$\begin{array}{ccc} \mathbf{\Sigma}'_1 & \xrightarrow{\sim} & \mathbf{\Sigma}'_2 \\ \downarrow & & \downarrow \\ \mathbf{\Sigma}_1 & \overset{\Phi}{\dashrightarrow} & \mathbf{\Sigma}_2 \end{array}$$

shows the congruence, where Φ is an affine isomorphism preserving the subsimplex structure. Such congruent families are maximal when the P_k are affinely independent.

Theorem 8.3 (Extension of Markov kernels to completed) *σ-algebras) Let $\Pi : \Omega \times \mathbf{S}' \to [0, 1]$ be a Markov kernel from $(\Omega, \mathbf{S})$ to $(\Omega', \mathbf{S}')$. For each $\omega \in \Omega$, let $(\Omega', (\mathbf{S}')^*_\omega, \Pi^*(\omega, \cdot))$ denote the completion of the probability space $(\Omega', \mathbf{S}', \Pi(\omega, \cdot))$.*

There exists a unique Markov kernel $\Pi^* : \Omega \times (\mathbf{S}')^* \to [0, 1]$ to the universal completion $(\Omega', (\mathbf{S}')^*)$ satisfying:

1. *for each $\omega \in \Omega$, $\Pi^*(\omega, \cdot)$ extends $\Pi(\omega, \cdot)$ to $(\mathbf{S}')^*$;*
2. *for each $A' \in (\mathbf{S}')^*$, $\omega \mapsto \Pi^*(\omega, A')$ is $\mathbf{S}^*$-measurable.*

Moreover, for any probability measure P on $(\Omega, \mathbf{S})$ with completion P^* on $(\Omega, \mathbf{S}^*)$, we have:

$$(P\Pi)^* = P^*\Pi^*$$

where:

- *$P\Pi$ is the measure on $(\Omega', \mathbf{S}')$ defined by $(P\Pi)(A') = \int_\Omega \Pi(\omega, A')P(d\omega)$;*
- *$(P\Pi)^*$ is its extension to $(\mathbf{S}')^*$;*
- *$P^*\Pi^*$ is defined by $(P^*\Pi^*)(B') = \int_\Omega \Pi^*(\omega, B')P^*(d\omega)$ for $B' \in (\mathbf{S}')^*$.*

Proof Step 1: Existence of Π^*

For each $\omega \in \Omega$, the probability measure $\Pi(\omega, \cdot)$ on $(\Omega', \mathbf{S}')$ extends uniquely to its completion

$$(\Omega', (\mathbf{S}')^*_\omega, \Pi(\omega, \cdot)^*).$$

The *universal completion* $(\mathbf{S}')^*$ is defined as:

$$(\mathbf{S}')^* = \bigcap_\mu (\mathbf{S}')^*_\mu,$$

where μ ranges over all finite measures on $(\Omega', \mathbf{S}')$.

Since $(\mathbf{S}')^* \subseteq (\mathbf{S}')^*_\omega$ for every ω (as $\Pi(\omega, \cdot)$ is finite), we define:

$$\Pi^*(\omega, A') = \Pi(\omega, \cdot)^*(A'), \quad \forall A' \in (\mathbf{S}')^*, \ \forall \omega \in \Omega.$$

This is well-defined because $\Pi(\omega, \cdot)^*$ extends $\Pi(\omega, \cdot)$ to $(\mathbf{S}')^*_\omega \supseteq (\mathbf{S}')^*$.
Step 2: Measurability of $\omega \mapsto \Pi^*(\omega, A')$

Fix $A' \in (\mathbf{S}')^*$. Define

$$f_{A'}(\omega) = \Pi^*(\omega, A').$$

We show $f_{A'}$ is $\mathbf{S}^*$-measurable.

Let

$$\mathcal{L} = \{A' \in (\mathbf{S}')^* \mid f_{A'} \text{ is } \mathbf{S}^*\text{-measurable}\}.$$

We have $\mathbf{S}' \subseteq \mathcal{L}$: If $A' \in \mathbf{S}'$, then $f_{A'}(\omega) = \Pi(\omega, A')$, which is $\mathbf{S}$-measurable by definition of a Markov kernel, hence also $\mathbf{S}^*$-measurable.

$\mathcal{L}$ is a σ-algebra:

- If $A' \in \mathcal{L}$, then $(A')^c \in \mathcal{L}$ because $f_{(A')^c}(\omega) = 1 - f_{A'}(\omega)$.
- If $\{A'_n\}_{n=1}^\infty \subset \mathcal{L}$, then $\bigcup_{n=1}^\infty A'_n \in \mathcal{L}$ because

$$f_{\bigcup_{n=1}^\infty A'_n}(\omega) = \lim_{m\to\infty} \Pi^*(\omega, \bigcup_{n=1}^m A'_n) = \lim_{m\to\infty} f_{\bigcup_{n=1}^m A'_n}(\omega),$$

and finite unions preserve measurability, limits preserve measurability.

Since $(\mathbf{S}')^*$ is the smallest σ-algebra containing $\mathbf{S}'$ and all universally measurable sets, we get $(\mathbf{S}')^* \subseteq \mathcal{L}$. Thus, $f_{A'}$ is $\mathbf{S}^*$-measurable for all $A' \in (\mathbf{S}')^*$
Step 3: Uniqueness of Π^*

Suppose $\widetilde{\Pi}^* : \Omega \times (\mathbf{S}')^* \to [0, 1]$ is another kernel with the same properties. For any $A' \in (\mathbf{S}')^*$ and $\omega \in \Omega$,

$$\widetilde{\Pi}^*(\omega, A') = \Pi^*(\omega, A'),$$

because both extend $\Pi(\omega, \cdot)$ uniquely from $\mathbf{S}'$ to $(\mathbf{S}')^*_\omega$.

Hence Π^* is unique.
Step 4: Commutation of Completions

Let P be a probability measure on $(\Omega, \mathbf{S})$, with completion P^* on $(\Omega, \mathbf{S}^*)$. Define:

$$(P\Pi)(A') = \int_\Omega \Pi(\omega, A')\, P(d\omega), \quad A' \in \mathbf{S}',$$

$$(P^*\Pi^*)(B') = \int_\Omega \Pi^*(\omega, B')\, P^*(d\omega), \quad B' \in (\mathbf{S}')^*.$$

We show $(P\Pi)^* = P^*\Pi^*$ on $(\mathbf{S}')^*$.

For $A' \in \mathbf{S}'$:

$$(P^*\Pi^*)(A') = \int_\Omega \Pi^*(\omega, A')\, P^*(d\omega) = \int_\Omega \Pi(\omega, A')\, P(d\omega) = (P\Pi)(A').$$

Thus both measures extend $P\Pi$ from $\mathbf{S}'$ to $(\mathbf{S}')^*$. By uniqueness of measure completion, $(P\Pi)^* = P^*\Pi^*$. The Markov kernel exists uniquely, satisfies the required measurability, and commutes with measure completions. □

Remark 8.1 We make a link between the statement outlined above and the universal property of Theorem 8.2 which sates that a **bijective correspondence** is established by the operators Π_{ij}, which are given by

$$\Pi_{ij} = \Pi^{(i)} \circ Q^{(j)},$$

where $\Pi^{(i)}$ is given by the mapping $\varphi_i(\omega)$.

Exercise 8.1 Another proof can be given using the definition of an inner and outer measure (that has not been introduced explicitly here). It is left as an exercise to reflect upon those different approaches.

8.2.5 Concluding with Klein–Plato's Framework

Corollary 8.1 *Let $\mathcal{F}_1 = \{P_\theta^{(1)}\}_{\theta\in\Theta}$ and $\mathcal{F}_2 = \{P_\theta^{(2)}\}_{\theta\in\Theta}$ be parametric families of probability measures on measurable spaces $(\Omega^{(1)}, \mathbf{S}^{(1)})$ and $(\Omega^{(2)}, \mathbf{S}^{(2)})$, respectively.*

- ***Figures:*** *$\mathcal{F}_1, \mathcal{F}_2$ are statistical analogues of Klein's geometric figures.*
- ***Fundamental Domain:*** *The pair $(\mathscr{E}, \mathscr{B})$ acts as a* geometric invariant, *analogous to a fundamental domain that characterizes congruence classes.*
- ***Motions:*** *The isomorphism ψ plays the role of a rigid motion, mapping the statistical figure $\mathcal{F}_1$ onto $\mathcal{F}_2$.*
- ***Kernels:*** *The Markov transition kernels $Q^{(i)}$ behave as probabilistic projections onto $\mathscr{E}$, generalizing the notion of quotient maps from classical geometry.*
- ***Fiber Condition:***

  $$Q^{(i)}(\varepsilon, \varphi_i^{-1}(\{\varepsilon\})) = 1,$$

  which ensures preservation of fibers (or orbits), reflecting the orbit-preserving nature of group actions in Klein's geometric framework.

Properties of statistical models are those invariant under the group of Markov morphisms.

Furthermore, when it comes to statistical Inference, hypothesis testing, estimation, and decision theory reduce to computing invariants on $\mathscr{E}$.

8.3 Markov Transformations and Associated Subcategories

The following theorem establishes the existence of a natural subcategory CAP* within the category CAP of measurable spaces and Markov morphisms.

Theorem 8.4 (Subcategory of Universally Complete Spaces) *Let* CAP *be the category where:*

- *Objects are measurable spaces* $(\Omega, \mathbf{S})$.
- *Morphisms are Markov kernels* $\Pi : \Omega \times \mathbf{S}' \to [0, 1]$.

The full subcategory CAP* *consists of objects* $(\Omega, \mathbf{S}^*)$ *where* $\mathbf{S}^*$ *is the universal completion of* $\mathbf{S}$[5].

There exists a functor $\mathscr{F}$: CAP → CAP* *defined by:*

$$\textit{Objects: } (\Omega, \mathbf{S}) \mapsto (\Omega, \mathbf{S}^*)$$
$$\textit{Morphisms: } \Pi \mapsto \Pi^*$$

where Π^* *is the unique extension to the universally completed* σ*-algebras.*

Remark 8.2 What role the universal completion of $\mathbf{S}$ plays in the theorem:

- **Kernel Extensions:** Markov kernels Π extend uniquely to $\widehat{\Pi}$ on $\mathbf{S}^*$ because:

  1. For each ω, $\Pi(\omega, \cdot)$ defines a measure.

[5] The universal completion $\mathbf{S}^*$ of a σ-algebra $\mathbf{S}$ (on a set Ω) is defined as $\mathbf{S}^* = \bigcap_\mu \mathbf{S}^*_\mu$, where:

- μ ranges over all finite measures on $(\Omega, \mathbf{S})$,
- $\mathbf{S}^*_\mu$ is the standard μ-completion of $\mathbf{S}$ (adding all μ-null sets).

- **Complete for Every Measure:** $\mathbf{S}^*$ contains all μ-null sets for every possible finite measure μ on $\mathbf{S}$ (unlike standard completions, which depend on a specific μ).
- **Minimal Extension:** $\mathbf{S}^*$ is the smallest σ-algebra containing $\mathbf{S}$ that remains complete under every finite measure.
- **Preserves Measurability:** any $\mathbf{S}$-measurable function remains $\mathbf{S}^*$-measurable.

2. $\mathbf{S}^*$ is complete with respect to all such measures simultaneously.

- **Functoriality:** the universal completion allows consistent extension of kernels across spaces:

$$\Pi : (\Omega_0, \mathbf{S}_0) \to (\Omega_1, \mathbf{S}_1) \quad \stackrel{*}{\to} \quad \Pi^* : (\Omega_0, \mathbf{S}_0^*) \to (\Omega_1, \mathbf{S}_1^*).$$

- **Category-Theoretic Significance:** the category CAP* (spaces with universally complete σ-algebras) is such that:
  - Morphisms compose without measurability issues.
  - Measure extensions behave functorially.
  - Probabilistic constructions become "universal."

Proof To verify CAP* is a subcategory:

1. **Object Inclusion**: By construction, all objects of CAP* are objects of CAP since $\mathbf{S}^* \supset \mathbf{S}$.
2. **Morphism Closure**: For any Markov kernel $\Pi : (\Omega_0, \mathbf{S}_0^*) \to (\Omega_1, \mathbf{S}_1^*)$, it remains a valid morphism in CAP as $\mathbf{S}_i^* \supset \mathbf{S}_i$.
3. **Identity Preservation**: The identity kernel $I(\omega, A) = \mathbf{1}_A(\omega)$ extends to $I^* = I$ on completed spaces.

For the functor $\mathscr{F}$:

- **Extension Uniqueness**: By Theorem 8.3, for any $\Pi : (\Omega_0, \mathbf{S}_0) \to (\Omega_1, \mathbf{S}_1)$, there exists a unique $\Pi^* : (\Omega_0, \mathbf{S}_0^*) \to (\Omega_1, \mathbf{S}_1^*)$ extending Π.
- **Composition Preservation**: For composable kernels:

$$\mathscr{F}(\Pi_{12} \circ \Pi_{01}) = (\Pi_{12} \circ \Pi_{01})^* = \Pi_{12}^* \circ \Pi_{01}^* = \mathscr{F}(\Pi_{12}) \circ \mathscr{F}(\Pi_{01})$$

The central equality follows from the universal property of completions and the measure-specific commutativity $(P\Pi)^* = P^*\Pi^*$ applied to each $P_\omega = \Pi_{01}(\omega, \cdot)$.

Thus $\mathscr{F}$ preserves identities and composition. □

Remark 8.3 The passage to the closure of the algebra in the category of measurable spaces is functorial.

We now show that Markov morphisms form a category in a natural way.

Lemma 8.1 (Measurable transformation under Markov kernels) *Let* $\Pi : \Omega \times \mathbf{S}' \to [0, 1]$ *be a Markov kernel from* $(\Omega, \mathbf{S})$ *to* $(\Omega', \mathbf{S}')$. *For any bounded measurable function* $f : \Omega' \to [0, 1]$ *that is* $\mathbf{S}'$*-measurable, the transformed function*

$$g(\omega) = (\Pi f)(\omega) = \int_{\Omega'} f(\omega')\Pi(\omega, d\omega')$$

is $\mathbf{S}$*-measurable and satisfies* $g(\omega) \in [0, 1]$ *for all* $\omega \in \Omega$.

Proof We establish both properties separately:

1. Measurability of g**:** Since f is $\mathbf{S}'$-measurable, the measurability of g follows from the definition of Markov kernels. Specifically:

- When $f = \mathbf{1}_A$ is an indicator function for $A \in \mathbf{S}'$, we have $g(\omega) = \Pi(\omega, A)$, which is $\mathbf{S}$-measurable by definition of Markov kernels.
- By linearity, measurability extends to simple functions (finite linear combinations of indicators).
- For general bounded measurable f, we use approximation: there exist simple functions $f_n \uparrow f$ with $0 \leq f_n \leq f \leq 1$, and by the Monotone Convergence Theorem:

$$g(\omega) = \int f d\Pi(\omega, \cdot) = \lim_{n\to\infty} \int f_n d\Pi(\omega, \cdot)$$

  Since each $\int f_n d\Pi(\omega, \cdot)$ is $\mathbf{S}$-measurable, and pointwise limits of measurable functions are measurable, g is $\mathbf{S}$-measurable.

2. Range Preservation $g(\omega) \in [0, 1]$**:** For each fixed ω, $\Pi(\omega, \cdot)$ is a probability measure. Thus:

$$\begin{aligned} 0 \leq g(\omega) &= \int_{\Omega'} f(\omega')\Pi(\omega, d\omega') \\ &\leq \int_{\Omega'} 1\Pi(\omega, d\omega') \quad (\text{since } f \leq 1) \\ &= \Pi(\omega, \Omega') = 1 \end{aligned}$$

Similarly, $g(\omega) \geq 0$ since $f \geq 0$. The bounds follow from the monotonicity of integration.
□

8.4 Dominated Families in the Category of Markov Morphisms

We now state a fundamental theorem concerning dominated families (denoted **DomFam**) in the context of Markov morphisms.

Theorem 8.5 (Invariance of Domination under Markov Morphisms) *Let μ and ν be measures on $(\Omega, \mathbf{S})$ with $\mu \gg \nu$ (i.e., μ dominates ν). For any Markov kernel $\Pi : \Omega \times \mathbf{S}' \to [0, 1]$ to $(\Omega', \mathbf{S}')$, the pushforward measures satisfy*

$$\mu\Pi \gg \nu\Pi.$$

That is, domination is preserved under Markov morphisms.

Proof Assume $\mu \gg \nu$ and fix $A \in \mathbf{S}'$ with $(\mu\Pi)(A) = 0$. Define

$$B = \{\omega \in \Omega : \Pi(\omega, A) > 0\}.$$

Since $\omega \mapsto \Pi(\omega, A)$ is $\mathbf{S}$-measurable (by definition of Markov kernel), $B \in \mathbf{S}$. By the non-negativity of $\Pi(\cdot, A)$:

$$\begin{aligned} 0 = (\mu\Pi)(A) &= \int\limits_{\Omega} \Pi(\omega, A)\mu(d\omega) \\ &\geq \int\limits_{B} \Pi(\omega, A)\mu(d\omega) \\ &\geq \mu(B) \cdot \inf_{\omega \in B} \Pi(\omega, A). \end{aligned}$$

Since $\Pi(\omega, A) > 0$ on B, we must have $\mu(B) = 0$. The domination $\mu \gg \nu$ implies $\nu(B) = 0$. Finally:

$$\begin{aligned} (\nu\Pi)(A) &= \int_B \Pi(\omega, A)\nu(d\omega) + \int_{\Omega \setminus B} \Pi(\omega, A)\nu(d\omega) \\ &= \int_B \Pi(\omega, A)\nu(d\omega) \quad (\text{since } \Pi(\omega, A) = 0 \text{ on } \Omega \setminus B) \\ &\leq \nu(B) = 0 \quad (\text{as } \Pi(\omega, A) \leq 1). \end{aligned}$$

Thus $(\nu\Pi)(A) = 0$, proving $\mu\Pi \gg \nu\Pi$. □

8.5 Dominated Families in the Category of Markov Morphisms

We will consider certain subcategories of CAP.
FAMD

- FAM: Stands for *family*, referring to the family of probability measures $\mathcal{P}$ associated with the measurable space $(\Omega, \mathbf{S})$.
- D: Stands for *dominated*, indicating that the family $\mathcal{P}$ is dominated by a σ-finite measure μ, i.e., $\mathcal{P} \ll \mu$.

Full Meaning: *Family of Absolutely Continuous Measures under Domination.*
FAMH

- FAM: Again stands for *family* (of probability measures $\mathcal{P}$).
- H: Stands for *homogeneous*, indicating that the measures in $\mathcal{P}$ are mutually absolutely continuous, i.e., $P_i \sim P_j$ for all $P_i, P_j \in \mathcal{P}$. This means they share the same null sets, reflecting a uniform "size" perception of measurable sets.

Full Meaning: *Family of Homogeneous Measures (in null sets).*

- FAMD: *Family + Domination .*
- FAMH: *Family + Homogeneity.*

Theorem 8.6 (Structure of Dominated Families) *Let* CAP *denote the category of measurable spaces with Markov morphisms.*

1. *The class of measurable spaces $(\Omega, \mathbf{S})$ equipped with a dominated family $\mathcal{P}$ of probability measures, together with all Markov morphisms between them, forms a full subcategory* FAMD *of* CAP.
2. *The class of spaces equipped with families of mutually absolutely continuous probability measures forms a full subcategory* FAMH *of* FAMD.

Proof **(i)** **FAMD is a subcategory:** Let $(\Omega, \mathbf{S}, \mathcal{P})$ be an object in FAMD where $\mathcal{P} = \{P_\theta\}_{\theta\in\Theta}$ is dominated by some Q. For any Markov kernel $\Pi : (\Omega, \mathbf{S}) \to (\Omega', \mathbf{S}')$, Theorem 8.5 implies:

$$Q\Pi \gg P_\theta \Pi \quad \forall\theta \in \Theta.$$

Thus $\{P_\theta \Pi\}_{\theta\in\Theta}$ is dominated in $(\Omega', \mathbf{S}')$. Composition of Markov kernels preserves this property, and identity morphisms trivially maintain domination.

(ii) FAMH is a subcategory: If $\mathcal{P}$ is mutually absolutely continuous ($P_i \sim P_j\ \forall i, j$), then for any $P_k \in \mathcal{P}$ and Markov kernel Π:

$$P_i\Pi \ll P_k\Pi \quad \text{and} \quad P_k\Pi \ll P_i\Pi \quad \forall i$$

again by Theorem 8.5. Thus $\{P_\theta\Pi\}$ remains mutually absolutely continuous. □

Remark 8.4 The universal completion functor $\mathscr{F}$: CAP → CAP* (from Theorem 8.4) preserves these subcategories.

Lemma 8.2 (Constructibility Under Domination) *Let P be a probability measure on (Ω, S) that is constructible (i.e., defined through an explicit Radon-Nikodym derivative dP/dQ relative to a reference measure Q). If P dominates a probability measure ν (i.e., $\nu \ll P$), then ν is also constructible.*

Proof Since $\nu \ll P$ and P is constructible, there exists a reference measure Q and $f_P = dP/dQ$. By the chain rule for Radon-Nikodym derivatives:

$$\frac{d\nu}{dQ} = \underbrace{\frac{d\nu}{dP}}_{\text{exists since } \nu \ll P} \cdot \underbrace{\frac{dP}{dQ}}_{f_P}.$$

Define $f_\nu = d\nu/dP$. Then $d\nu/dQ = f_\nu \cdot f_P$ is explicitly constructible from f_P and f_ν. Thus ν is constructible relative to Q. □

8.5.1 Conclusion and Remarks

Dominated families and mutually absolutely continuous families form full subcategories FAMD and FAMH of CAP, with structure preserved under Markov morphisms and universal completion. Furthermore we have:

$$\text{FAMH} \subset \text{FAMD} \subset \text{CAP}.$$

△ Note that the subcategory FAMD is useful in statistical models since it generates a parametric model $\{P_\theta\}_{\theta\in\Theta}$ where for instance θ can be the mean of a Gaussian distributions.

8.5.2 Dominating Measures and Constructive Probabilities

Lemma 8.3 (Finite Mixture Dominance) *For probability measures $P_1, \ldots, P_n$ on (Ω, S), the mixture*

$$P_0(\cdot) = \frac{1}{n} \sum_{k=1}^{n} P_k(\cdot)$$

dominates each P_k.

Exercise 8.2 Show that for each P_k:

$$P_k(\cdot) \leq n \cdot P_0(\cdot)$$

8.6 Homogeneous Klein Geometries and Geodesic Structures

Definition 8.2 A Klein geometry (G, H) with associated homogeneous space $X = G/H$ is *quasi-homogeneous* if for any two points $x, y \in X$ and any open neighborhood $U_y \subset X$ of y, there exists $g \in G$ such that:

$$g \cdot x \in U_y.$$

This means the G-action is *topologically almost transitive*.

Lemma 8.4 *In a quasi-homogeneous Klein geometry* (G, H) *with* $X = G/H$ *connected, every continuous* G*-invariant function* $f : X \to \mathbb{R}$ *(i.e.,* $f(g \cdot x) = f(x)\ \forall g \in G, \forall x \in X$*) is constant.*

Proof Assume f is continuous and G-invariant. Fix $x_0 \in X$ and let $c = f(x_0)$. For any $y \in X$ and $\epsilon > 0$, consider the open ball $B_\epsilon(c) \subset \mathbb{R}$. By continuity, $f^{-1}(B_\epsilon(c))$ is open in X.

Since X is quasi-homogeneous, there exists $g \in G$ such that $g \cdot y \in f^{-1}(B_\epsilon(c))$. By G-invariance:

$$f(y) = f(g \cdot y) \in B_\epsilon(c)$$

Thus $|f(y) - c| < \epsilon$ for all $\epsilon > 0$, so $f(y) = c$. As y was arbitrary, f is constant. □

8.6.1 Geodesic Structures

Definition 8.3 Let $(\mathcal{M}, \nabla)$ be a manifold with affine connection, and $\gamma : I \to \mathcal{M}$ a smooth curve parametrized by $t \in I \subseteq \mathbb{R}$. The curve γ is a *geodesic* if its tangent vector field $X(t) = \dot{\gamma}(t)$ is parallel along γ, i.e.,

$$\nabla_{\dot{\gamma}} \dot{\gamma} = 0 \quad \text{for all } t \in I.$$

Remark 8.5 Geodesics generalize straight lines in Euclidean geometry to curved spaces, representing paths of"zero acceleration."

In local coordinates $(x^1, \ldots, x^n)$, the geodesic equation is:

$$\frac{d^2x^k}{dt^2} + \sum_{i,j=1}^{n} \Gamma^k_{ij} \frac{dx^i}{dt} \frac{dx^j}{dt} = 0, \quad k = 1, \ldots, n$$

The affine parameterization is canonical: if $s = s(t)$ is another parameterization under which γ remains geodesic, then $s(t) = at + b$ for constants $a \neq 0, b$.

8.6.2 Torsion, Curvature, and Flat Connections

An affine connection ∇ defines two fundamental tensors characterizing local geometry:

- *Torsion Tensor* $\mathrm{T} : \mathfrak{X}(\mathcal{M}) \times \mathfrak{X}(\mathcal{M}) \to \mathfrak{X}(\mathcal{M})$:

$$\mathrm{T}(X, Y) = \nabla_X Y - \nabla_Y X - [X, Y]$$

- *Riemann Curvature Tensor* $R : \mathfrak{X}(\mathcal{M}) \times \mathfrak{X}(\mathcal{M}) \times \mathfrak{X}(\mathcal{M}) \to \mathfrak{X}(\mathcal{M})$:

$$R(X, Y)Z = \nabla_X \nabla_Y Z - \nabla_Y \nabla_X Z - \nabla_{[X,Y]} Z$$

where $[X, Y] = XY - YX$ is the Lie bracket of vector fields.

Remark 8.6 When $\mathrm{T} \equiv 0$ and $R \equiv 0$:

- The connection is *flat* (integrable).
- $(\mathcal{M}, \nabla)$ is locally isomorphic to affine space $\mathbb{R}^n$.
- Parallel transport depends only on the homotopy class of paths (relative to endpoints).
- There exist local coordinates where $\Gamma^k_{ij} = 0$ and geodesics are affine lines.

8.6.3 Totally Geodesic Submanifolds

Definition 8.4 A submanifold $\mathcal{N}$ of a manifold $\mathcal{M}$ equipped with an affine connection ∇ is *totally geodesic* if every geodesic in $\mathcal{M}$ that is tangent to $\mathcal{N}$ at some point remains entirely within $\mathcal{N}$. More precisely:

$$\gamma(0) \in \mathcal{N} \text{ and } \dot{\gamma}(0) \in \mathcal{T}_{\gamma(0)}(\mathcal{N}) \quad \Longrightarrow \quad \gamma(t) \in \mathcal{N} \text{ for all } t$$

Lemma 8.5 *A submanifold $\mathcal{N} \subset \mathcal{M}$ is totally geodesic if and only if for any vector fields X, Y tangent to $\mathcal{N}$, the covariant derivative $\nabla_X Y$ is also tangent to $\mathcal{N}$.*

Proof ($\Rightarrow$) Suppose $\mathcal{N}$ is totally geodesic. For $p \in \mathcal{N}$ and $X_p, Y_p \in \mathcal{T}_p(\mathcal{N})$, extend to vector fields $X, Y \in \mathfrak{X}(\mathcal{N})$. Let $\gamma(t)$ be the geodesic in $\mathcal{M}$ with $\gamma(0) = p$ and $\dot{\gamma}(0) = X_p$. Since $\mathcal{N}$ is totally geodesic, $\gamma \subset \mathcal{N}$ and $\nabla_{\dot{\gamma}}\dot{\gamma} = 0$. Consider the vector field Y along γ. By the geodesic equation and compatibility:

$$\nabla_X Y|_p = \lim_{t \to 0} \frac{P_t(Y_{\gamma(t)}) - Y_p}{t}$$

where P_t is parallel transport along γ. Since $\gamma \subset \mathcal{N}$ and Y is tangent to $\mathcal{N}$, both $Y_{\gamma(t)}$ and its parallel transport remain in $\mathcal{T}(\mathcal{N})$. Thus $\nabla_X Y|_p \in \mathcal{T}_p(\mathcal{N})$.

($\Leftarrow$) Assume $\nabla_X Y$ tangent to $\mathcal{N}$ for X, Y tangent to $\mathcal{N}$. Let γ be a geodesic in $\mathcal{M}$ with $\gamma(0) \in \mathcal{N}$ and $\dot{\gamma}(0) \in \mathcal{T}_{\gamma(0)}(\mathcal{N})$. Since $\nabla_{\dot{\gamma}}\dot{\gamma} = 0$ and $\dot{\gamma}$ is initially tangent to $\mathcal{N}$, the condition implies $\dot{\gamma}$ remains tangent to $\mathcal{N}$. Thus γ stays in $\mathcal{N}$. □

Remark 8.7 For Riemannian manifolds $(\mathcal{M}, g)$ with the Levi-Civita connection:

- $\mathcal{N}$ is totally geodesic iff the second fundamental form vanishes.
- Geodesics in $\mathcal{N}$ (with induced metric) coincide with geodesics in $\mathcal{M}$.
- "Locally shortest" curves in $\mathcal{N}$ are geodesics in both $\mathcal{N}$ and $\mathcal{M}$.

8.7 Equivariant Connections in the Category of Geometric Structures

Let **Geo** be a category of smooth manifolds with affine connections, where morphisms are smooth maps preserving connection structures.

Definition 8.5 Let ∇ be an affine connection on objects of **Geo**. The connection is *equivariant* if for every morphism $\varphi : (\mathcal{M}, \nabla^{\mathcal{M}}) \to (\mathcal{N}, \nabla^{\mathcal{N}})$ in **Geo**, φ maps geodesics to geodesics. Specifically, for any geodesic γ in $\mathcal{M}$:

$$\varphi \circ \gamma \text{ is a geodesic in } \mathcal{N}.$$

This categorical equivariance ensures geodesic structures are preserved under admissible transformations.

8.7.1 Geodesic Midpoints and Natural Transformations

The midpoint operation induces a natural transformation in equivariant categories:

- For any geodesic segment $\overline{pq}$ in $(\mathcal{M}, \nabla^{\mathcal{M}})$, there exists a unique *midpoint* $m(p, q)$ satisfying:
$$d(p, m) = d(m, q) \quad \text{(in affine parameter)},$$
- The midpoint function $m : \mathcal{M} \times \mathcal{M} \to \mathcal{M}$ is symmetric: $m(p, q) = m(q, p)$,
- For equivariant connections, midpoints transform naturally:
$$\varphi(m(p, q)) = m(\varphi(p), \varphi(q)) \quad \forall \varphi \in \mathrm{Mor}(\mathbf{Geo}).$$

Theorem 8.7 (Covariance of Midpoints) *If* ∇ *is equivariant in* **Geo**, *then for all morphisms* $\varphi : \mathcal{M} \to \mathcal{N}$ *and pairs* $(p, q) \in \mathcal{M} \times \mathcal{M}$:

$$\begin{array}{ccc} \mathcal{M} \times \mathcal{M} & \xrightarrow{m_{\mathcal{M}}} & \mathcal{M} \\ {\scriptstyle \varphi\times\varphi}\big\downarrow & & \big\downarrow{\scriptstyle \varphi} \\ \mathcal{N} \times \mathcal{N} & \xrightarrow[m_{\mathcal{N}}]{} & \mathcal{N} \end{array}$$

commutes. This naturality makes m a covariant functor under **Geo**.

Proof Let γ be the geodesic from p to q with midpoint $m_{\mathcal{M}}(p, q)$. Since $\varphi \circ \gamma$ is geodesic in $\mathcal{N}$ (by equivariance), its midpoint is $m_{\mathcal{N}}(\varphi(p), \varphi(q))$. But $(\varphi \circ \gamma)(t_{\mathrm{mid}}) = \varphi(m_{\mathcal{M}}(p, q))$, hence:
$$\varphi(m_{\mathcal{M}}(p, q)) = m_{\mathcal{N}}(\varphi(p), \varphi(q)).$$
□

Remark 8.8 This categorical perspective highlights several conceptual insights:

- **Affine Symmetries**: Equivariant connections preserve geodesics under morphisms, encoding a functorial version of affine symmetry.
- **Symmetric Space Flavor**: The midpoint operation equips $(\mathcal{M}, \nabla^{\mathcal{M}})$ with a reflection symmetry reminiscent of classical symmetric spaces, now phrased in categorical terms.

- **Erlangen Program Revisited**: Naturality of midpoints extends Klein's Erlangen program from transformation groups acting on spaces to functorial connections acting on categories.
- **Functorial Geometry**: Geodesic structures become objects of a functorial calculus, where curvature and symmetry can be studied through categorical morphisms rather than local coordinates.

Categorical and Geometric Structures in Statistical Manifold Theory 9

In the previous chapter, we introduced key ideas in probability and statistics and briefly alluded to a more structural perspective using category theory and differential geometry. However, these tools have not yet been systematically employed. In this chapter, we take a decisive step in that direction by developing a structured approach integrating category theory and differential geometry for the study of statistical manifolds.

The categorical viewpoint allows us to formalize probabilistic transformations, treating probability distributions as objects and Markov kernels as morphisms. Differential geometry provides powerful tools for analyzing the intrinsic structure of statistical manifolds through connections, curvature, and geodesics. By synthesizing these perspectives, we gain deeper insight into the geometric foundations of statistical models and their transformations.

This chapter establishes these connections explicitly. We first develop the categorical framework for statistical manifolds, then demonstrate how geometric structures emerge naturally within this formalism. This synthesis clarifies fundamental relationships while enabling new developments in information geometry, learning theory, and related fields.

9.1 Foundational Collections of Probability Measures

To rigorously define statistical manifolds, we introduce five fundamental collections of probability measures[1]. These serve as categorical building blocks when equipped with appropriate structures:

[1] The terms "probability measure" and "probability distribution" are often used interchangeably, but subtle distinctions exist depending on context. We use "probability measures" for defining $\mathrm{Cap}(\Omega, \mathbf{S})$ to emphasize that the set is defined via axioms of measure theory and that there is no reliance on topologies, densities, or random variables. Probability distributions refer to a broader term that may

© The Author(s), under exclusive license to Springer Nature Switzerland AG 2026

N. C. Combe et al., *Exploring Information Geometry*, Frontiers in Mathematics,
https://doi.org/10.1007/978-3-032-23054-6_9

The following abbreviations are firmly grounded in the theory of capacities, as evidenced by the literature cited: [Chq54, DM75, KaW96, Me68].

Definition 9.1 (*Canonical Probability Collections*) Let $(\Omega, \mathbf{S})$ be a measurable space and $\mathbf{Z} \subset \mathbf{S}$ a σ-ideal of null sets. We define:

- $\mathsf{Cap}(\Omega, \mathbf{S})$: All probability measures on $(\Omega, \mathbf{S})$.
  - **Origin:** Abbreviation of "capacity" (French: *capacité*), rooted in Choquet capacity theory, see [Chq54].
  - **Context:** In geometric measure theory, capacities generalize measures to characterize the "size" of sets. Here, it denotes the set of all probability measures, aligning with the idea of capacities assigning a "probabilistic size" to measurable sets.
- $\mathsf{Capd}(\Omega, \mathbf{S}, \mathbf{Z})$: Probability measures absolutely continuous w.r.t. a fixed reference measure μ where $\mathbf{Z} = \{A \in \mathbf{S} : \mu(A) = 0\}$.
  - **Origin:** "Capacity with dominance" (or "dominated capacities").
  - **Context:** The subscript d signifies domination by a reference measure μ. Measures here are absolutely continuous with respect to μ, reflecting the Radon–Nikodym property (existence of densities). Dominated families are central in statistics (e.g., Hellinger metrics).
- $\mathsf{Caph}(\Omega, \mathbf{S}, \mathbf{Z})$: Probability measures with *exactly* $\mathbf{Z}$ as their null sets, i.e., $P(A) = 0 \iff A \in \mathbf{Z}$.
  - **Origin:** "Capacity homogeneous in null sets."
  - **Context:** The subscript h denotes homogeneity, meaning all measures share the same σ-ideal of null sets Z. This is an equivalence class under mutual absolute continuity (e.g., equivalent measures in a statistical model).
- $\mathsf{Conh}(\Omega, \mathbf{S}, \mathbf{Z})$: Families of mutually absolutely continuous probability measures sharing $\mathbf{Z}$ as common null sets.
  - **Origin:** "Capacity homogeneous in null sets."
  - **Context:** The subscript h denotes homogeneity, meaning all measures share the same σ-ideal of null sets Z. This is an equivalence class under mutual absolute continuity (e.g., equivalent measures in a statistical model).

---

refer to: probability measures (in abstract spaces)/ cumulative distribution functions (CDFs) for $\mathbb{R}^n$ or density/mass functions (PDFs/PMFs) under a reference measure.

- $\mathsf{Var}(\Omega, \mathbf{S})$: Finite signed measures of bounded variation.
  - **Origin:** Standard abbreviation for "variation" (as in bounded variation).
  - **Context:** Refers to signed measures with finite total variation:

$$\|\mu\|_{\mathrm{TV}} < \infty.$$

Remark 9.1 The σ-ideal $\mathbf{Z}$ typically represents topologically negligible sets or null sets of a distinguished measure. For Caph, measures are equivalent to μ ($P \sim \mu$), while for Conh, all measures in a family are mutually equivalent.

Notation	Definition/requirement	Example/interpretation
$\mathsf{Capd}(\Omega, \mathbf{S}, \mathbf{Z})$	Requires absolute continuity w.r.t. a fixed reference measure μ. Measures may have more null sets than $\mathbf{Z}$.	If μ is Lebesgue measure on $\mathbb{R}$, Capd includes all measures with densities w.r.t. Lebesgue measure.
$\mathsf{Caph}(\Omega, \mathbf{S}, \mathbf{Z})$	Requires exact equivalence to μ: $P(A) = 0 \iff A \in \mathbf{Z}$ (identical null sets). All measures share the same σ-ideal $\mathbf{Z}$.	For μ the Lebesgue measure, Caph is the singleton class of measures equivalent to μ.
$\mathsf{Conh}(\Omega, \mathbf{S}, \mathbf{Z})$	A set of families of measures, each family consisting of measures mutually absolutely continuous and sharing $\mathbf{Z}$.	Facilitates studying global properties (e.g., convexity) of sets of equivalent measure classes; typically a singleton for fixed $\mathbf{Z}$.

The relationship between Cap and Var is characterized by:

Proposition 9.1 (Measure Correspondence) *There exists a canonical embedding (but not a bijection) from the set of probability measures into the vector space of signed measures of bounded variation:*

$$\iota : \mathsf{Cap}(\Omega, \mathbf{S}) \hookrightarrow \mathsf{Var}(\Omega, \mathbf{S}).$$

Proof The inclusion map $\iota : \mathsf{Cap} \hookrightarrow \mathsf{Var}$, $\iota(P) = P$ is canonical. Every probability measure is a signed measure with $P(\Omega) = 1$ and $P \geq 0$, so ι is well-defined.

However, signed measures with negative values or total variation $\neq 1$ exist in Var but not in Cap.

For instance, there exist $\nu \in \mathsf{Var}$ with $\nu \notin \mathsf{Cap}$, such as:

- $\nu = -\delta_\omega$ (negative values);
- $\nu = 2\delta_\omega$ (total mass $\neq 1$).

Thus ι is not bijective. □

9.2 Categorical Framework for Statistical Manifolds

These collections form the basis for categorical structures, where:

- **Objects**: elements of the classes Cap, Capd, Caph, Conh, Var;
- **Morphisms**: Markov kernels $\Pi : (\Omega_1, \mathbf{S}_1) \to (\Omega_2, \mathbf{S}_2)$.

The commutative diagram below illustrates the *functorial* relationships:

$$\begin{array}{ccc} \mathsf{Capd} & \hookrightarrow & \mathsf{Cap} \\ {\scriptstyle T_\Pi}\downarrow & & \downarrow{\scriptstyle T_\Pi} \\ \mathsf{Capd}_\Pi & \hookrightarrow & \mathsf{Cap}_\Pi \end{array},$$

where T_Π denotes pushforward by Markov kernel Π and Cap_Π (resp. Capd_Π) corresponds to pushforward measures on $(\Omega_2, \mathbf{S}_2)$.

An additional layer in this context is introduced by the **geometric structure**, namely connections and metrics arising from statistical divergences.

Note that the geometric structures are compatible with pushforwards:

$$g_{\Pi_P}(d\Pi X, d\Pi Y) = g_P(X, Y)$$

for tangent vectors X, Y.

Proposition 9.2 *The spaces admit manifold structures as follows:*

- Var *is a Banach manifold.*
- Caph *is a smooth manifold.*
- Cap *and* Capd *are convex sets with manifold boundaries.*
- Conh *is a union of convex manifolds (one per equivalence class)*[2]

To summarize we can write things as follows:

$\triangle$ This has statistical and geometric implications. Manifold operations (geodesics, divergences) are only well-defined on Caph and boundary points in Cap/Capd correspond to *degenerate models*. In particular, tangent spaces are undefined at boundary points and the curvature blows up near boundaries.

[2] Conh consists of equivalence classes of mutually absolutely continuous probability measures sharing the same null sets $\mathbf{Z}$. Each equivalence class $[P] = Q : Q \sim P$ (where $Q \sim P$ means $Q \ll P$ and $P \ll Q$) forms: a convex set and a smooth manifold. But Conh as a whole is not a manifold because its components are disconnected.

Space	Convex	Manifold
Var	Yes	Banach manifold
Caph	Yes	Smooth manifold
Cap	Yes	Boundary only
Capd	Yes	Boundary only
Conh	No	Per-class manifold

Proof See Chentsov [Che82], pp. 73–74. □

A statistical problem is naturally associated with the measurable space $(\Omega, \mathbf{S})$ together with additional data such as:

- **Model**: A family $\{P_\theta\}_{\theta\in\Theta}$ on $(\Omega, \mathbf{S})$;
- **Parameter Space**: Measurable space $(\Theta, \boldsymbol{T})$;
- **Prior**: A probability measure π on $(\Theta, \boldsymbol{T})$ (Bayesian) or constraint set $\Theta_0 \subset \Theta$ (frequentist);
- **Decision Space**: Measurable space $(\Omega', \mathbf{S}')$ of *actions*[3];
- **Loss**: Measurable function $L : \Theta \times \Omega' \to \mathbb{R}$.[4]
- **Decision Rule**: Markov kernel $\Pi : \Omega \times \mathbf{S}' \to [0, 1]$.

We give a complete definition of the latter.

Definition 9.2 A *statistical decision rule* is a Markov kernel $\Pi : \Omega \times \mathbf{S}' \to [0, 1]$ where:

- $\Pi(\omega, \cdot)$ is a probability measure on $(\Omega', \mathbf{S}')$ $\forall \omega \in \Omega$.
- $\Pi(\cdot, A')$ is $\mathbf{S}$-measurable $\forall A' \in \mathbf{S}'$.

Definition 9.3 (*Categories of Statistical Decisions*) The categories CAP, CAPD, CAPH, CONH have respectively:

- **Objects**: Triples $(\Omega, \mathbf{S}, \mathscr{C})$ where $\mathscr{C}$ is one of the following: $\mathsf{Cap}(\Omega, \mathbf{S})$, $\mathsf{Capd}(\Omega, \mathbf{S}, \mathbf{Z})$, $\mathsf{Caph}(\Omega, \mathbf{S}, \mathbf{Z})$, $\mathsf{Conh}(\Omega, \mathbf{S}, \mathbf{Z})$.
- **Morphisms**: Markov kernels $\Pi : (\Omega_1, \mathbf{S}_1) \times \mathbf{S}_2 \to [0, 1]$ satisfying:

$$P \in \mathscr{C}_1 \implies P\Pi \in \mathscr{C}_2.$$

[3] Actions refer to the set of possible choices or decisions available to a statistician or decision-maker after observing data. For example in medical testing, actions could be {treat, do not treat}.

[4] Loss function connection: $L(\theta, a)$ quantifies the penalty for choosing action a when θ is the true parameter.

Composition is given by:

$$(\Pi_1 \circ \Pi_2)(\omega, A'') = \int_{\Omega_2} \Pi_2(\omega_2, A'')\Pi_1(\omega, d\omega_2).$$

Proof The proof follows from the categorical structure imposed by the Markov homeomorphisms and the natural compatibility of the decision rule formulation. □

Remark 9.2 The category defined by CAPD is a full subcategory of FAMD (see previous chapter), but not equal to it. The category defined by CAPH is a full subcategory of FAMH, but not equal to it. The discrepancy arises because FAMD and FAMH allow arbitrary families of measures, while CAPD and CAPH restrict to maximal families induced by a σ-ideal $\mathbf{Z}$.

- Dominated families (FAMD) generalize CAPD to arbitrary dominated sets.
- Homogeneous families (FAMH) generalize CAPH to arbitrary mutually absolutely continuous sets.

△ Note also that the CAPD generate all models dominated by μ (e.g., nonparametric density estimation in contrast with what we saw for FAMD).

9.2.1 Types of Manifolds in Theory of Statistical Manifolds

9.2.1.1 Geometric Structure of Conh The Simplex Case

Consider the category CAPH_F (where the index F in CAPH_F stands for finite) of spaces $\mathsf{Caph}(\Omega, \mathbf{S}, \mathbf{Z})$ where the quotient σ-algebra $\mathbf{S}/\mathbf{Z}$ is finite. These spaces form finite-dimensional manifolds. Any measure μ on $(\Omega, \mathbf{S})$ with null ideal $\mathbf{Z}$ is completely determined by its values on the atoms of $\mathbf{S}/\mathbf{Z}$.[5]

Definition 9.4 (*Natural Simplex Structure*) Let $\mathsf{Conh}(\Omega, \mathbf{S}, \mathbf{Z})$ be the set of all positive measures on $(\Omega, \mathbf{S})$ that vanish exactly on $\mathbf{Z}$ (i.e., $\mu(A) = 0 \iff A \in \mathbf{Z}$). If $\mathbf{S}/\mathbf{Z}$ has m atoms $A_1, \ldots, A_m$, then the correspondence

$$\mu \longleftrightarrow \boldsymbol{\mu} = (\mu(A_1), \mu(A_2), \cdots, \mu(A_m)) = (\mu_1, \mu_2, \cdots, \mu_m)$$

with $\mu_j > 0$ defines a global chart. This is called the *natural chart* of Conh.

[5] Caph with finite $\mathbf{S}/\mathbf{Z}$ forces measures to be discrete on finitely many atoms, yielding a finite-dimensional parametric model.

The manifold structure includes other charts defined by diffeomorphic coordinate transformations, but the natural chart is privileged through its direct measure-theoretic interpretation.

Proposition 9.3 **(Probability Simplex as Hypersurface)** *The space* $\mathsf{Caph}(\Omega, \mathbf{S}, \mathbf{Z})$ *embeds into* $\mathsf{Conh}(\Omega, \mathbf{S}, \mathbf{Z})$ *as the hypersurface defined by the normalization condition*

$$\langle \boldsymbol{\mu}, \mathbf{1} \rangle = \sum_{j=1}^{m} \mu_j = 1.$$

Specifically, Caph *is the intersection:*

$$\mathsf{Caph} = \mathsf{Conh} \cap \left\{ \boldsymbol{\mu} \in \mathbb{R}^m : \sum \mu_j = 1, \mu_j > 0 \right\}.$$

The probability vectors $\mathbf{p} = (P(A_1), \ldots, P(A_m))$ *parameterize the interior of the standard* $(m-1)$*-simplex:*

$$\Delta^{m-1} = \left\{ (p_1, \ldots, p_m) : p_j > 0, \sum p_j = 1 \right\}$$

with vertices $\mathbf{e}_j$ *(standard basis vectors in* $\mathbb{R}^m$*).*

Definition 9.5 (*Dual Coordinate Systems*) For a probability measure $P \in \mathsf{Caph}(\Omega, \mathbf{S}, \mathbf{Z})$ with atomic σ-algebra $\mathbf{S}/\mathbf{Z} = \{A_1, \ldots, A_m\}$, the values $p_j = P(A_j)$ simultaneously serve as:

- **Natural Coordinates**: The bijection

$$P \longleftrightarrow \mathbf{p} = (p_1, \ldots, p_m)$$

  identifies points in the ambient space $\mathbb{R}^m$.
- **Barycentric Coordinates**: The affine representation

$$P = \sum_{j=1}^{m} p_j \delta_{A_j}$$

  where δ_{A_j} are Dirac measures at the atoms, forming vertices of the simplex.

9.2.2 Redundancy in Natural Coordinates

Proposition 9.4 *The natural coordinates* $\mathbf{p} = (p_1, \ldots, p_m)$ *are linearly dependent as coordinates on* Caph*, reducing the intrinsic dimension to* $m-1$.

Proof The proof is evident. The constraint $\sum_{j=1}^{m} p_j = 1$ implies:

$$\text{rank}\left(\frac{\partial(p_1, \ldots, p_m)}{\partial(\xi_1, \ldots, \xi_{m-1})}\right) = m - 1$$

where ξ are independent coordinates. The Jacobian has a null space spanned by $(1, 1, \ldots, 1)^T$, confirming dependency. □

9.2.3 Invariant Local Coordinate Systems

To define intrinsic coordinates, we fix an arbitrary index ordering and express:

$$p_m = 1 - \sum_{j=1}^{m-1} p_j, \quad p_j > 0, \quad \sum_{j=1}^{m-1} p_j < 1$$

using coordinates $\mathbf{p} = (p_1, \ldots, p_{m-1})$. However, this breaks permutation symmetry because such charts lack invariance under permutations of the atoms.

To resolve this, we introduce the following:

1. Homogeneous Coordinates.

To achieve permutation invariance, we represent measures in the projective space via equivalence classes under scaling:

$$[\mathbf{P}] = \{c \cdot \mathbf{p} \mid c > 0\}, \quad \mathbf{p} = (p_1, \ldots, p_m)$$

where two vectors are equivalent if they differ by a positive scalar multiple. This homogeneous representation has interesting properties since this:

- eliminates the normalization constraint $\sum p_j = 1$;
- makes coordinates invariant under atom relabeling;
- For a finite sample space with m atoms, Conh embeds as an open dense submanifold in $\mathbb{RP}^{m-1}$ via:

$$(p_1, \ldots, p_m) \mapsto [p_1 : \cdots : p_m],$$

  where $[p_1 : \cdots : p_m]$ are homogeneous coordinates.

2. Rectilinear Coordinates

Another type of coordinates, which can be used to solve the problem of chart invariance, is to introduce the rectilinear coordinates. For intrinsic parameterization independent of atom ordering, define rectilinear coordinates via expectations:

Definition 9.6 (*Rectilinear Coordinate System*) Let $\{f_j\}_{j=0}^{m-1}$ where $f_j : \Omega \to \mathbf{R}$ are linearly independent **S**-measurable functions with $f_0 \equiv 1$. For $P \in \mathsf{Caph}$, the rectilinear coordinates are:

$$t^j(P) = \mathbb{E}_P[f_j] = \int_\Omega f_j(\omega) P(d\omega), \quad 1 \le j \le m-1$$

provided the matrix $\left(f_j(\omega_i)\right)$ has full rank m at sample points $\{\omega_i\}_{i=1}^m$ and f is integrable with respect to the measure P (or quasi-integrable).

Remark 9.3 Some key properties:

1. $t^0(P) = \mathbb{E}_P[1] = 1$ (normalization);
2. The coordinate map $P \mapsto \mathbf{t} = (t^1, \ldots, t^{m-1})$ is a diffeomorphism;

Example 9.1 A coordinate system $\boldsymbol{\eta} = (\eta^1, \ldots, \eta^{m-1})$ on Caph is natural if:

$$\eta^j(P) = \mathbb{E}_P[\phi_j] = \int_\Omega \phi_j(\omega) P(d\omega),$$

where $\{\phi_j\}$ are linearly independent functionals. Each ϕ_j is P-quasi-integrable and induces a Fisher-orthogonal frame: $g_{ij} = \mathrm{cov}_P(\phi_i, \phi_j)$.

In particular, the following are equivalent, in the context of exponential families:

1. $\boldsymbol{\eta}$ are natural coordinates.
2. The Fisher metric $g_{ij} = \partial_i \partial_j \psi(\boldsymbol{\eta})$ is Hessian.
3. Geodesics are straight lines in $\boldsymbol{\eta}$-coordinates.

Precisely, we can apply this for the exponential family $dP_\eta = \exp\left(\sum_{j=1}^{m-1} \eta^j \phi_j - \psi(\boldsymbol{\eta})\right) d\mu$ where we deduce that:

- $\boldsymbol{\eta}$ are natural coordinates.
- $\mathbb{E}_{P_\eta}[\phi_j] = \partial_j \psi(\boldsymbol{\eta})$.
- Dual coordinates given by $\boldsymbol{\theta} = \nabla \psi(\boldsymbol{\eta})$ are expectation parameters.

9.2.4 Canonical Affine Coordinates in the Theory of Statistical Manifolds

The introduction of canonical affine coordinates in statistical manifold theory arises naturally from fundamental geometric and probabilistic considerations. Consider a measurable space $(\Omega, \mathbf{S})$ partitioned into m disjoint atoms $\{A_1, \ldots, A_m\}$. Define indicator functions:

$$\chi_j(\omega) = \begin{cases} 1 & \text{if } \omega \in A_j \\ 0 & \text{otherwise} \end{cases}$$

While $\{\chi_1, \ldots, \chi_m\}$ are linearly dependent ($\sum_{j=1}^{m} \chi_j = 1$), any $m-1$ of them together with the constant function $\mathbf{1}(\omega) \equiv 1$ form a linearly independent set.

Definition 9.7 A coordinate system $\mathbf{s} = (s^1, \ldots, s^{m-1})$ on the open simplex corresponding to $\mathsf{Caph}(\Omega, \mathbf{S}, \mathbf{Z})$ is *canonical affine* if

$$P_{\mathbf{s}}(\cdot) = \exp\left[\sum_{\alpha=1}^{m-1} s^\alpha g_\alpha(\cdot) - \Psi(\mathbf{s})\right] P_0(\cdot),$$

where $P_{\mathbf{s}}$ is the probability measure at coordinate point $\mathbf{s}$ on the open simplex, P_0 is the base measure (origin), $\{g_\alpha\}_{\alpha=1}^{m-1}$ are coordinate functions, and $\mathbf{s}$ is a covariant linear coordinate system of the group of translations.

Explicitly, if P_0 represents the probability distribution at the origin and $g_0(\omega) \equiv 1$ and $\{g_1, \ldots, g_{m-1}\}$ are linearly independent, the Radon–Nikodym derivative with respect to a dominating measure μ is then:

$$\frac{dP_{\mathbf{s}}}{d\mu}(\omega) = p(\omega; \mathbf{s}) = p(\omega; 0) \exp\left[\sum_{\alpha=1}^{m-1} s^\alpha g_\alpha(\omega) - \Psi(\mathbf{s})\right]$$

where μ is a $\mathbf{Z}$-positive measure vanishing only on $\mathbf{Z}$-sets, $p(\omega; 0) = \frac{\mathrm{d}P_0}{\mathrm{d}\mu}$ is the density of the base measure P_0 and the log-partition function $\Psi(\mathbf{s})$ is:

$$\exp[\Psi(\mathbf{s})] = \int_\Omega \exp\left[\sum_{\alpha=1}^{m-1} s^\alpha g_\alpha(\omega)\right] dP_0(\omega).$$

9.2.4.1 Tangent Vectors and Tangent Spaces

To analyze the structure of statistical manifolds, we introduce tangent vectors and vector fields. Let $\{P_t\}_{t \in I}$ be a smooth curve of probability measures parameterized by t in an interval $I \subseteq \mathbb{R}$. The *tangent vector* to the curve at $P = P_\theta$ (where $\theta \in I$) is defined as the derivation:

$$X_P(f) = \frac{d}{dt} f(P_t)\bigg|_{t=\theta}$$

for smooth functions f on the manifold.

Lemma 9.1 *Consider a statistical manifold with finite sample space $\Omega = \{A_1, \ldots, A_m\}$.*

- *The curve P_t admits the first-order expansion:*

$$P_t(\cdot) = P_\theta(\cdot) + \tau\nu(\cdot) + o(\tau)$$

where $\tau = t - \theta$, ν is a signed measure satisfying $\nu(\Omega) = 0$, and the norm of the remainder term vanishes faster than τ.

- *There exists a linear isomorphism between the tangent space at P_θ and the space of signed measures ν with $\nu(\Omega) = 0$.*

Proof Represent P_t by its probability vector $\mathbf{p}(t) = (p_1(t), \ldots, p_m(t))$ where $p_j(t) = P_t(A_j)$. Near $t = \theta$, expand:

$$p_j(t) = p_j(\theta) + \tau\dot{p}_j(\theta) + o(\tau), \quad j = 1, \ldots, m-1$$

The constraint $\sum_{j=1}^m p_j(t) = 1$ implies:

$$p_m(t) = 1 - \sum_{j=1}^{m-1} p_j(t) = p_m(\theta) - \tau\sum_{j=1}^{m-1} \dot{p}_j(\theta) + o(\tau)$$

Define the signed measure ν by $\nu(A_j) = \dot{p}_j(\theta)$ for $j = 1, \ldots, m-1$ and $\nu(A_m) = -\sum_{j=1}^{m-1} \dot{p}_j(\theta)$, ensuring $\nu(\Omega) = 0$. For any $H \subseteq \Omega$:

$$P_t(H) = \sum_{A_j \subseteq H} p_j(t) = P_\theta(H) + \tau\nu(H) + o(\tau)$$

(with uniform convergence in total variation norm). For a smooth function f:

$$\frac{d}{dt} f(\mathbf{p}(t))\bigg|_{t=\theta} = \sum_{j=1}^{m-1} \frac{\partial f}{\partial p_j}\dot{p}_j(\theta) = \sum_{j=1}^{m} \frac{\partial f}{\partial p_j}\nu(A_j)$$

establishing the isomorphism. The linearity and continuity follow from the finite-dimensional structure. □

9.2.5 Ceva Lines and Tangent Vector Fields

In the probability simplex Δ^{m-1}, a *Ceva line*[6] from vertex $\mathbf{e}_j$ to a point $\mathbf{q}$ on the opposite face is defined as:

$$\mathbf{p}^{(j)}(t) = p_j(t)\mathbf{e}_j + [1 - p_j(t)]\mathbf{q}, \quad \text{where} \quad \mathbf{q} = \sum_{i \neq j} q_i \mathbf{e}_i, \quad \sum_{i \neq j} q_i = 1.$$

Lemma 9.2 *For the Ceva line from* $\mathbf{e}_1$*, there exists a parameterization where the coordinate* $p_1(t)$ *satisfies the differential equation:*

$$\frac{dp_1}{dt} = p_1(1 - p_1)$$

with explicit solution:

$$p_1(t) = \frac{1}{1 + e^{-t}}.$$

The corresponding tangent vector field is $X_1 = \frac{\partial}{\partial t}$.

Remark 9.4 The vector field X_1 generates the logistic parameterization. An alternative parameterization Y_1 uses:

$$p_1(t) = 1 - e^{-t}.$$

For a general vertex $\mathbf{e}_j$, the Ceva line is:

$$\mathbf{p}^{(j)}(t) = p_j(t)\mathbf{e}_j + [1 - p_j(t)] \sum_{i \neq j} q_i \mathbf{e}_i.$$

The tangent field X_j satisfies $\dot{p}_j = p_j(1 - p_j)$, while Y_j satisfies $p_j(t) = 1 - e^{-t}$.

Proof For the parameterization ($j = 1$):

$$p_1(t) = (1 + e^{-t})^{-1}$$

$$\frac{dp_1}{dt} = \frac{e^{-t}}{(1 + e^{-t})^2} = \frac{1}{1 + e^{-t}} \cdot \frac{e^{-t}}{1 + e^{-t}} = p_1(1 - p_1).$$

For $i \neq 1$, the components are $p_i(t) = [1 - p_1(t)]q_i$, so:

$$\frac{dp_i}{dt} = -\frac{dp_1}{dt} q_i = -p_1(1 - p_1)q_i = -p_1 p_i.$$

[6] We refer here to the Ceva and Menelaus theorems, fundamental in Euclidean geometry.

The tangent vector to the curve is:

$$\dot{\mathbf{p}}(t) = \dot{p}_1\mathbf{e}_1 + \sum_{i=2}^{m} \dot{p}_i\mathbf{e}_i = p_1(1-p_1)\mathbf{e}_1 - p_1\sum_{i=2}^{m} p_i\mathbf{e}_i.$$

Noting that $\mathbf{p}(t) = p_1\mathbf{e}_1 + \sum_{i=2}^{m} p_i\mathbf{e}_i$, we can write:

$$\dot{\mathbf{p}}(t) = p_1\left(\mathbf{e}_1 - \mathbf{p}(t)\right).$$

Thus the tangent field is proportional to $\mathbf{e}_1 - \mathbf{p}(t)$. □

Lemma 9.3 *The system of tangent vectors $\{X_j\}_\mathbf{p}$ for $j = 1, \ldots, m$ is complete at each point $\mathbf{p}$ in the probability simplex Δ^{m-1}. That is:*

1. *Any tangent vector $Z_\mathbf{p}$ can be expressed as a linear combination of the $\{X_j\}_\mathbf{p}$.*
2. *Any $m-1$ vectors from $\{X_j\}_\mathbf{p}$ form a basis for the tangent space $\mathcal{T}_\mathbf{p}(\Delta^{m-1})$.*

Moreover, every tangent vector field Z admits a unique expansion:

$$Z = \sum_{j=1}^{m} \zeta^j(\mathbf{p})X_j$$

subject to the constraint:

$$\sum_{j=1}^{m} \zeta^j(\mathbf{p})p_j = 0.$$

All alternative expansions $Z = \sum_{j=1}^{m} \eta^j(\mathbf{p})X_j$ satisfy:

$$\eta^j(\mathbf{p}) = \zeta^j(\mathbf{p}) + \phi(\mathbf{p}) \quad \text{for all } j = 1, \ldots, m$$

where $\phi(\mathbf{p})$ is a scalar function.

Remark 9.5 The lemma remains valid if we replace the constraint $\sum_{j=1}^{m} \zeta^j p_j = 0$ with:

$$\sum_{j=1}^{m} \zeta^j = 0.$$

Proof Let $Z_\mathbf{p}$ correspond to a signed measure μ with $\mu(\Omega) = 0$. Representing μ in the standard basis:

$$\mu = \sum_{j=1}^{m} \mu_j\delta_j \quad \text{where} \quad \sum_{j=1}^{m} \mu_j = 0.$$

Since X_j corresponds to the tangent vector $p_j(\mathbf{e}_j - \mathbf{p})$, we have:

$$Z_{\mathbf{p}} = \sum_{j=1}^{m} \frac{\mu_j}{p_j}(X_j)_{\mathbf{p}}.$$

This gives an expansion $Z = \sum_{j=1}^{m} \zeta^j X_j$ with $\zeta^j = \mu_j / p_j$. The constraint follows from:

$$\sum_{j=1}^{m} \zeta^j p_j = \sum_{j=1}^{m} \mu_j = 0.$$

The linear dependence relation $\sum_{j=1}^{m} X_j = \mathbf{0}$ implies that any $m - 1$ vectors span the $(m-1)$-dimensional tangent space. For uniqueness, consider two expansions:

$$\sum_{j=1}^{m} \zeta^j X_j = Z = \sum_{j=1}^{m} \eta^j X_j.$$

Their difference satisfies:

$$\sum_{j=1}^{m} (\zeta^j - \eta^j) X_j = \mathbf{0}.$$

Since the null space is one-dimensional, $\zeta^j - \eta^j = \phi(\mathbf{p})$ for some scalar function ϕ. The constraint determines ϕ uniquely:

$$\sum_{j=1}^{m} \zeta^j p_j = \sum_{j=1}^{m} (\eta^j + \phi) p_j = 0 + \phi \sum_{j=1}^{m} p_j = \phi,$$

so $\phi = \sum_{j=1}^{m} \zeta^j p_j = 0$ for the constrained expansion. □

9.2.6 Affine Connections and Ceva Lines on Statistical Manifolds

We now examine the relationship between canonical affine coordinates and Ceva lines in the probability simplex Δ.

Lemma 9.4 *In canonical affine coordinates with basis functions $g_\alpha = \chi_\alpha$ for $\alpha \neq k$, the coordinate curves defined by varying a single coordinate:*

$$x^i = t, \quad x^j = \text{constant for } j \neq i$$

correspond to Ceva lines of the i-th family. The tangent vector fields to these curves are:

$$X_i = \frac{\partial}{\partial x^i}.$$

Proof Consider the canonical affine coordinate system with base measure P_0 at the simplex center:

$$P_0(A_j) = \frac{1}{m} \quad \text{for } j = 1, \ldots, m.$$

The probability density is:

$$\frac{dP_{\mathbf{x}}}{d\mu}(\omega) = p(\omega; \mathbf{x}) = p(\omega; 0) \exp\left[\sum_{\alpha=1}^{m-1} x^\alpha g_\alpha(\omega) - \Psi(\mathbf{x})\right],$$

where $\{g_\alpha\}_{\alpha=1}^{m-1}$ are the coordinate functions. For $g_\alpha = \chi_\alpha$ (indicator of atom A_α), integration over A_i yields:

$$p_i(\mathbf{x}) = P_{\mathbf{x}}(A_i) = \frac{1}{m} \exp\left[x^i - \Psi(\mathbf{x})\right] \quad \text{for } i = 1, \ldots, m-1$$
$$p_m(\mathbf{x}) = P_{\mathbf{x}}(A_m) = \frac{1}{m} \exp\left[-\Psi(\mathbf{x})\right],$$

with the partition function:

$$\exp[\Psi(\mathbf{x})] = \frac{1}{m}\left(1 + \sum_{\alpha=1}^{m-1} \exp[x^\alpha]\right).$$

Consider a curve where only x^1 varies: $x^1(t) = x_0^1 + t$ while $x^\alpha = x_0^\alpha$ for $\alpha = 2, \ldots, m-1$. The probabilities become:

$$p_1(t) = \frac{1}{m} \exp\left[x_0^1 + t - \Psi(t)\right]$$
$$p_j(t) = \frac{1}{m} \exp\left[x_0^j - \Psi(t)\right] \quad \text{for } j = 2, \ldots, m-1$$
$$p_m(t) = \frac{1}{m} \exp\left[-\Psi(t)\right],$$

where $\Psi(t) = \Psi(x_0^1 + t, x_0^2, \ldots, x_0^{m-1})$. The ratio of probabilities for $j \geq 2$ remains constant:

$$\frac{p_j(t)}{p_k(t)} = \frac{\exp[x_0^j]}{\exp[x_0^k]} = \text{constant}.$$

This implies the curve is a Ceva line:

$$\mathbf{p}(t) = p_1(t)\mathbf{e}_1 + [1 - p_1(t)] \sum_{j=2}^{m} q_j \mathbf{e}_j,$$

where $q_j = \frac{\exp[x_0^j]}{\sum_{k=2}^m \exp[x_0^k]}$ for $j \geq 2$.

Differentiating $p_1(t)$ and using the partition function constraint $\sum_{j=1}^{m} p_j(t) = 1$:

$$\frac{dp_1}{dt} = p_1(t) - p_1(t)^2 = p_1(t)(1 - p_1(t)),$$

confirming the tangent field $X_1 = \frac{\partial}{\partial x^1}$ generates the required result. The general case follows by permutation of indices. □

9.3 Translation-Invariant Tensor Fields

Theorem 9.1 *A tensor field on the probability simplex is invariant under the group of affine translations if and only if its components are constant in canonical affine coordinates.*

Proof Consider canonical affine coordinates $\mathbf{z} = (z^1, \ldots, z^{m-1})$. In these coordinates, Δ^{m-1} is an $(m-1)$-dimensional affine space where translations act as parallel shifts: $\mathrm{T}_\mathbf{s} : \mathbf{z} \mapsto \mathbf{z} + \mathbf{s}$.

First, for an invariant scalar field f:

$$f(\mathrm{T}_\mathbf{s}\mathbf{p}) = f(\mathbf{p}) \quad \forall \mathbf{s}$$

Since translations are transitive, f must be constant: $f(\mathbf{p}) = c$ for all $\mathbf{p} \in \Delta^{m-1}$.

For a vector field $Y = \sum_{j=1}^{m-1} a^j(\mathbf{p}) \frac{\partial}{\partial z^j}$:

$$\mathrm{T}_\mathbf{s}\left(\left.\frac{\partial}{\partial z^j}\right|_\mathbf{p}\right) = \left.\frac{\partial}{\partial z^j}\right|_{\mathrm{T}_\mathbf{s}\mathbf{p}}$$

$$Y_\mathbf{p} = \mathrm{T}_\mathbf{s}\left(Y_{\mathrm{T}_\mathbf{s}^{-1}\mathbf{p}}\right)$$

Invariance implies:

$$\sum_j a^j(\mathbf{p}) \left.\frac{\partial}{\partial z^j}\right|_\mathbf{p} = \sum_j a^j(\mathrm{T}_\mathbf{s}\mathbf{p}) \left.\frac{\partial}{\partial z^j}\right|_{\mathrm{T}_\mathbf{s}\mathbf{p}}$$

Since $\left\{\frac{\partial}{\partial z^j}\right\}$ form a basis, $a^j(\mathbf{p}) = a^j(\mathrm{T}_\mathbf{s}\mathbf{p})$ for all $\mathbf{s}$, so a^j are constant.

The general tensor case follows similarly: tensor components must be constant as basis tensors transform invariantly. □

Corollary 9.1 *There exist translation-invariant Riemannian metrics on the simplex that make it isometric to Euclidean space* $\mathbb{R}^{m-1}$.

9.4 Flat Connections

Definition 9.8 An affine connection ∇ is *flat* if it satisfies:

$$\nabla_{X_j} X_k = 0 \quad \forall j, k = 1, \ldots, m$$

where $\{X_j\}$ are the canonical vector fields.

Remark 9.6 A flat connection is compatible with any translation-invariant Riemannian metric g:

$$\nabla g = 0.$$

9.5 Exercises

Some easy exercises.

- Exercise 1. Determine the affine transformation which maps the points $(0, 0)$, $(1, 0)$ and $(0, 1)$ to the points
  - $(0, -1)$, $(1, 1)$, and $(-1, 1)$, respectively,
  - $(-4, -5)$, $(1, 7)$, and $(2, -9)$, respectively.
- Exercise 2.The triangle $\triangle ABC$ has vertices $A(2, 0)$, $B(-3, 0)$ and $C(3, -3)$, and the points $P(-1, -1)$, $Q(1, 3)$ and $R(-\frac{1}{4}, 0)$ lie on BC, CA, and AB respectively.
  - Determine the ratios in which P, Q, and R divide the sides of the triangle.
  - Determine whether or not the points P, Q,and R are collinear.
- Exercise 3. Consider a statistical manifolds of exponential type. Using the charts description proof that the structure of statistical manifold is affine and flat.

9.5.1 Some Corrections

Exercise 8.1, correction.

Proof The proof is divided into three parts.

First Part of the Proof

Firstly, since any measure on **6** can be extended to a measure on $\mathbf{6}^*$, where it coincides with the induced inner and outer measures, it remains to show that for any fixed $A \in \mathbf{6}^*$, the transition probability function $\omega \mapsto \Pi^*(\omega, A)$ is $\mathbf{S}^*$-measurable.

Second Part of the Proof

Having established the first part of the proof, we now proceed to the second step. Our aim is to show that the real-valued function

$$\Pi_A(\omega) = \Pi(\omega; A)$$

is $\mathbf{S}^*$-measurable for any fixed $A \in \mathbf{6}^*$. To do so, it suffices to verify that for any probability measure P on $\mathbf{S}$ and any real number $z \in [0, 1]$, the outer and inner P-measures of the preimage

$$U = \{\omega \mid \Pi_A(\omega) < z\}$$

of the half-open interval $[0, z)$ coincide, i.e.,

$$\overline{P}[U] = \underline{P}[U],$$

which ensures that U is indeed $\mathbf{S}^*$-measurable.

Consider probability measures P and Q. For any $A \in \mathbf{6}^*$, there exist two sets $\mathcal{G}, \mathcal{F} \in \mathbf{6}$ satisfying:

$$\mathcal{F} \subset A \subset \mathcal{G},$$

such that for every ω, the following equalities hold:

$$Q[\mathcal{G}] = \overline{Q}[A] = \underline{Q}[A] = Q[\mathcal{F}].$$

Since probability measures are monotone, we obtain the inequalities:

$$\Pi(\omega; \mathcal{G}) \geq \Pi^*(\omega; A) \geq \Pi(\omega; \mathcal{F}).$$

Therefore, the functions $\Pi(\omega; \mathcal{G})$ and $\Pi(\omega; \mathcal{F})$ are $\mathbf{S}$-measurable and equal almost everywhere, since

$$\int \Pi(\omega; \mathcal{G}) P\{d\omega\} = \int \Pi(\omega; \mathcal{F}) P\{d\omega\},$$

except possibly on a $\mathbf{S}$-measurable P-null set N where $P\{N\} = 0$.

From this, it follows that:

$$\{\omega \mid \Pi(\omega; \mathcal{G}) < z\} \subseteq \{\omega \mid \Pi^*(\omega; A) < z\}.$$

Moreover, we also have:

$$\{\omega \mid \Pi^*(\omega; A) < z\} \subseteq \{\omega \mid \Pi(\omega; \mathcal{G}) < z\} \cup N.$$

Thus, we conclude:

$$\overline{P}[U] = \underline{P}\{\omega \mid \Pi(\omega; \mathcal{G}) < z\},$$

which proves the second part.

Third Part of the Proof

Finally, we proceed to the third and last step: proving that the measures $(P\Pi)^*$ and $P^*\Pi^*$, both defined on $(\Omega', \mathbf{6}^*)$, coincide. That is, we claim that for any $A \in \mathbf{6}^*$, we have:

$$(P\Pi)^*\{A\} = (P^*\Pi^*)\{A\}.$$

Let us choose sets $\mathcal{F}$ and $\mathcal{G}$ satisfying for a given A:

$$\mathcal{F} \subseteq A \subseteq \mathcal{G}.$$

Then, as before, we have:

$$Q\{\mathcal{G}\} = Q^*\{A\} = Q\{\mathcal{F}\},$$

where $Q = P\Pi$. Given that the function $\Pi^*_A(\omega) = \Pi^*(\omega; A)$ is $\mathbf{6}^*$-measurable, we obtain:

$$(P\Pi)\{\mathcal{G}\} = \int \Pi(\omega; \mathcal{G}) P[d\omega] = \int \Pi^*(\omega; \mathcal{G}) P^*[d\omega].$$

By monotonicity, we conclude:

$$\int \Pi^*(\omega; \mathcal{G}) P^*[d\omega] \geq \int \Pi^*(\omega; A) P^*[d\omega].$$

Thus, we deduce:

$$(P^*\Pi^*)\{A\} = (P\Pi)\{\mathcal{G}\}.$$

Since this holds for all $A \in \mathbf{6}^*$, we obtain the desired equality:

$$(P^*\Pi^*)\{A\} = (P\Pi)^*\{A\}.$$

This completes the proof.

□

Exercise 8.2.
The measure P_0 is dominating since for each P_k we have the bound

$$P_k\{.\} \leq 2^k P_0\{.\}$$

and the inclusion of null sets:

$$\mathbf{Z}_{P_0} \subset \mathbf{Z}_{P_k}.$$

If we can show that P_0 is a constructive probability measure, the desired conclusion follows from Lemma 8.2.

Consider a measurable mapping $\omega = f_k(x)$, where $f_k : \mathbf{E} \to \Omega$ determines the constructive distribution

$$P_k\{.\} = \lambda \frac{1}{f_k(.)}.$$

Define f_0 piecewise as follows: for $2^{-k} < x < 2^{-k+1}$, let

$$f_0(x) = g_k(x) = f_k(2^k x - 1),$$

and set $f_0(0) = f_1(0)$.

This function is measurable, as it is composed of countably many measurable functions, each defined on a distinct measurable subset. Consequently, the f_0-preimage of any $\mathbf{S}$-set is a countable union of disjoint g_k-preimages.

Furthermore, we have the measure transformation:

$$\lambda g_k^{-1}\{.\} = 2^{-1} \lambda f_k^{-1}\{.\},$$

since the g_k-preimage is the f_k-preimage shifted by 2^{-k} and contracted by a factor of 2^k. This completes the proof.

Part III

Frobenius Manifolds and Information Geometry

Frobenius Manifolds 10

Part III operates at a more advanced level than the preceding sections, as it presumes familiarity with fundamental concepts of algebraic topology–such as connected sums, Chern numbers, Homology, Klein bottles, Möbius strips, etc. We draw on these notions to illuminate ideas that, while sometimes overlooked in recent literature, remain essential–particularly the concepts of affine structures and affine manifolds, which play a central role in the theory of Frobenius manifolds.

10.1 A Short Story of Frobenius Manifolds

A Frobenius manifold $\mathcal{M}$ is a geometric realization of a certain class of partial nonlinear differential equations of degree 3. These equations, commonly referred to as the WDVV equations–named after Witten, Dijkgraaf, Verlinde, and Verlinde–are also known in some mathematical contexts as the Associativity Equations.

Notice that the notion of Frobenius manifolds represents a geometrical object, whereas the WDVV equations belong to the realm of analysis. The terminology “Frobenius manifolds” and “Associativity equations” appears predominantly in mathematical literature; “WDVV equations” is more frequently encountered in discussions closer to physics.

These perspectives, however, are deeply interwoven. As shown by Yu. I. Manin [Ma99] and B. Dubrovin [D96], they ultimately coincide.

In the context of physics, solutions of the WDVV equations encode the moduli spaces of topological conformal field theories. These solutions are pivotal in the formulation of mirror

© The Author(s), under exclusive license to Springer Nature Switzerland AG 2026

N. C. Combe et al., *Exploring Information Geometry*, Frontiers in Mathematics,
https://doi.org/10.1007/978-3-032-23054-6_10

symmetry for Calabi–Yau 3-folds. Notably, specific solutions to the WDVV equations with particular properties serve as generating functions for the Gromov–Witten invariants of Kähler or symplectic manifolds.

10.1.1 Homological Mirror Symmetry (and More!)

In the 1990s, the concept of Frobenius manifolds gained traction among both algebraic and analytic geometers due to the discovery of three significant classes, all stemming from the profound interplay between mathematics and physics. This development led to the creation of advanced mathematical tools and generated an abundance of challenging and intriguing problems. Some of these arose naturally from attempts to provide a rigorous mathematical interpretation of mirror symmetry (e.g., Homological Mirror Symmetry). Others emerged from the ambition to axiomatize and deepen the mathematical understanding of topological quantum field theory.

The first class of Frobenius manifolds originated in the study of singularities. Specifically, Saito spaces–predating the explicit definition of Frobenius manifolds–were associated with the unfolding spaces of singularities. A particularly straightforward example is the space of complex polynomials of a fixed degree d with distinct roots, which can also be identified with the configuration space of d marked points on the complex plane.

Subsequent classes discovered in the 1990s have a formal nature. Examples include the (formal) moduli spaces of solutions to Maurer–Cartan equations modulo gauge equivalence and the formal completions of cohomology spaces of smooth projective (or compact symplectic) manifolds. The latter is commonly referred to as quantum cohomology [KonMa94].

More recently, a new class of Frobenius manifolds has emerged, demonstrating the versatility of this structure. It has been shown that under certain conditions, the manifold of probability distributions can generate a Frobenius manifold (this is the Combe–Manin construction [CM20]). This discovery inspires many open questions, particularly regarding the complete classification of Frobenius manifolds, the identification of novel classes, and the elucidation of the intricate relationships between the existing ones.

In the first section of this chapter, we provide a concise survey of Frobenius manifolds. We introduce the definitions of Frobenius, pre-Frobenius, and potential pre-Frobenius manifolds and briefly outline the WDVV equations. We also touch upon the hidden algebraic structures intrinsic to Frobenius manifolds. This section concludes with a brief summary.

In the second section, we delve more deeply into the theoretical framework of Frobenius manifolds.

In the last (third) section, we present a collection of exercises and open problems for the reader to explore.

10.2 Preliminary Notions: To Be (or Not) Affine

A cornerstone in the theory of Frobenius manifolds is the presence of an affine flat structure, often simply referred to as an affine structure. Without this geometric framework, the very notion of a Frobenius manifold cannot be formulated. This naturally leads us to a fundamental question: **when does a manifold admit an affine structure**, and is there a meaningful way to classify all such manifolds? We answer this question in this section.

Let $\mathcal{M}$ be a smooth manifold of dimension n. An affine structure on an n-dimensional manifold $\mathcal{M}$ is defined via a collection of coordinate charts $\{(U_\alpha, \phi_\alpha)\}$, where $\{U_\alpha\}$ forms an open cover of $\mathcal{M}$, and $\phi_\alpha : U_\alpha \to \mathbb{R}^n$ is a local coordinate system such that the transition functions $\phi_\beta \circ \phi_\alpha^{-1}$ are affine transformations on $\phi_\alpha(U_\alpha \cap U_\beta)$, mapping it to $\phi_\beta(U_\alpha \cap U_\beta)$.

Recall that the group of affine transformations is given by:

$$\left\{ \begin{pmatrix} A & b \\ 0 & 1 \end{pmatrix} \;\middle|\; A \in GL(n, \mathbb{R}),\ b \in \mathbb{R}^n \right\}.$$

Definition 10.1 An affine manifold is a smooth manifold equipped with an affine structure.

The existence of an affine flat structure on $\mathcal{M}$ is equivalent to the presence of a specific class of connections on the tangent bundle of $\mathcal{M}$. Namely, there is a bijective correspondence between affine flat structures and flat, torsion-free affine connections ∇ on $\mathcal{M}$.

In the context of differential geometry, a manifold over a field $\mathbb{K}$ (real or complex) with an affine structure is characterized by a tangent bundle whose underlying G-structure corresponds to the group of affine transformations $Aff(n) = GL(n, \mathbb{K}) \rtimes \mathbb{K}^n$, where $GL(n, \mathbb{K})$ is the general linear group over $\mathbb{K}$.

Alternatively, an affine flat structure can be described by a subsheaf $\mathcal{T}(\mathcal{M})^f \subset \mathcal{T}(\mathcal{M})$ of linear spaces of pairwise commuting vector fields. Locally, one has the tensor product over the ground field:

$$\mathcal{T}(\mathcal{M}) = \mathcal{O}_\mathcal{M} \otimes \mathcal{T}(\mathcal{M})^f.$$

Sections of $\mathcal{T}(\mathcal{M})^f$ correspond to flat vector fields. Furthermore, the metric g is compatible with the structure $\mathcal{T}(\mathcal{M})^f$ if $g(X, Y)$ is constant for all flat vector fields X and Y (see previous chapters).

Are there richer or more exotic examples of affine manifolds beyond the familiar Euclidean spaces and tori? In what follows, we explore examples of affine manifolds that go beyond the classical cases of Euclidean space and the torus. This naturally leads us to consider crystallographic groups and, in particular, Bieberbach groups. The following part denotes a range of related results that can be tied to the Chern Conjecture. Rather than delving into technical proofs, our goal here is to survey some of the most compelling developments. However, we warn that the reader might need some background in algebraic topology.

10.2.1 Crystallographic Groups and Fundamental Groups

Crystallographic groups are indispensable for quantifying order in atomic lattices and cosmic large-scale structures. More mathematically, concerning flat manifolds, crystallographic groups classify compact flat Riemannian manifolds (via Bieberbach theorems [Bb11, Bb12]). The fundamental group of a compact, complete, flat affine manifold is known as an *affine crystallographic group*. This means it is a discrete subgroup

$$\Gamma \subseteq \mathrm{Aff}(\mathbb{R}^n) = \mathbb{R}^n \rtimes \mathrm{GL}(n, \mathbb{R})$$

that acts properly discontinuously and cocompactly on $\mathbb{R}^n$, and the manifold itself emerges as the quotient $\mathbb{R}^n / \Gamma$.

A natural question arises: *Can such an affine crystallographic group contain torsion?* In other words, do there exist nontrivial finite-order elements in Γ?

In the Euclidean setting, Bieberbach's theorem guarantees that a crystallographic group of isometries is virtually torsion-free–that is, it contains a torsion-free subgroup of finite index. But this property fails in the broader affine context: affine crystallographic groups *can* indeed have torsion. A simple example is:

$$\Gamma = \{(z, \pm I_n) \mid z \in \mathbb{Z}^n\} \subset \mathrm{Aff}(\mathbb{R}^n),$$

which is discrete and cocompact. However, it is *not* torsion-free since the reflection element $(0, -I_n)$ has order 2. This makes Γ a valid affine crystallographic group that is not torsion-free.

The following property based on the literature given in [Aus64, FG83, G22] outlines the relation between compact, complete flat affine manifolds and affine crystallographic groups:

Lemma 10.1 *The fundamental group of a compact, complete flat affine manifold is an affine crystallographic group.*

A Bieberbach group is a crystallographic being torsion-free (contains no non-identity elements of finite order). A **Bieberbach group** Γ satisfies therefore the following properties:

1. **Free Action**: Γ acts freely (fixed-point-free) on $\mathbb{R}^n$.
   This implies that the quotient $\mathcal{M} = \mathbb{R}^n / \Gamma$ is a compact flat Riemannian manifold (and that it has no singularities).
2. **Structure**: This group satisfies the short exact sequence:

$$0 \to \mathbb{Z}^n \to \Gamma \to G \to 1,$$

   where:

   - $\mathbb{Z}^n$ is the maximal abelian subgroup (i.e., lattice of translations).
   - G is a finite group (the **holonomy group**), with $G \hookrightarrow GL(n, \mathbb{Z})$.

3. **Holonomy**: the holonomy group G is isomorphic to $\Gamma/\mathbb{Z}^n$ and classifies the topology of $\mathcal{M}$.

The quotient $\mathbb{R}^n / \Gamma$ is a compact flat manifold. Bieberbach classified crystallographic groups for which the quotient $\mathbb{R}^n / \Gamma$ is compact (and up to affine equivalence—equivalence under conjugation by affine transformations). This resolved the first part of Hilbert's 18th problem, showing that in each dimension, only finitely many compact flat manifolds exist up to affine equivalence. For example,

- $n = 2$: the torus $\mathbb{T}^2$.
- $n = 3$: 10 isomorphism classes (e.g., didicosm, tricosm).

For further reading on the subject, we suggest the recent book on the geometry of crystallographic groups by A. Szczepański: [Szcz24].

> The Klein-Plato framework (or Kleinian approach to geometry) centers on classifying geometric structures through transformation groups. Bieberbach groups fit naturally into this framework, particularly in the context of Euclidean structures.

10.2.2 2-Dimensional Affine Structures

Let us consider two well-known structures in topology: the torus and the Klein bottle. These are the only compact 2-dimensional manifolds that admit **affine** structures. Here, "affine" refers to a structure defined by an atlas of charts to $\mathbb{R}^2$ with affine transition maps. For closed surfaces, affine structures coincide with flat Riemannian (Euclidean) structures. For more details, we refer to [Bnz60].

Assume $\mathcal{M}$ is a closed manifold distinct from the torus or Klein bottle. Then there exists *no* affine structure on it. This follows directly from:

Theorem 10.1 (Benzécri, [Bnz60]) *A closed surface admits an affine structure if and only if its Euler characteristic vanishes.*[1]

10.2.3 Affine Structures on Complex Surfaces

For complex manifolds, Kobayashi [Ko79] established a complete classification of compact complex surfaces admitting holomorphic affine connections:

Theorem 10.2 (Kobayashi, [Ko79]) *A compact complex surface admits a holomorphic affine connection if and only if it is one of the following:*

1. *Complex tori;*
2. *Hyperelliptic surfaces;*
3. *Minimal elliptic surfaces with odd first Betti number (b_1 odd), positive geometric genus ($p_g > 0$), and vanishing Chern number squared ($c_1^2 = 0$);*
4. *Minimal surfaces with $b_1 = 1$, $b_2 = 0$, and $p_g = 0$, excluding Hopf surfaces covered by primary Hopf surfaces satisfying:*

$$\pi_1(\mathcal{M}) \cong \mathbb{Z} \ltimes \mathbb{Z}[1/p], \quad p \in \mathbb{Z}_{\geq 1}$$

Remarkably, Kobayashi proved that *no other* compact complex surfaces admit holomorphic affine connections.

10.2.4 n-Dimensional Affine Structures, $n \geq 3$

For manifolds of dimension $n \geq 3$, there is no complete characterization for the existence of affine structures.

Smillie's theorem provides a key obstruction:

[1] The Euler characteristic χ is a topological invariant. For polyhedral surfaces, $\chi = V - E + F$ (vertices/edges/faces). A tetrahedron has $\chi = 4 - 6 + 4 = 2$, and a cube has $\chi = 8 - 12 + 6 = 2$. Homology generalizes χ to higher dimensions. Closed surfaces with $\chi = 0$ (torus, Klein bottle) admit affine structures; others (e.g., $\chi = 2$ for S^2) do not.

Theorem 10.3 (Smillie, [Sm81]) *Let $\mathcal{M}$ be a compact connected manifold of positive dimension. If $\pi_1(\mathcal{M})$ is built out of finite groups by taking free products, direct products and finite extensions, then M does not admit an affine structure.*

For example, consider the connected sum $\mathcal{M}\#\mathcal{N}$ of two closed n-manifolds ($n \geq 3$) where each has fundamental group isomorphic to $\mathbb{Z}/3\mathbb{Z}$ (e.g., lens spaces $L(3, 1)$). The fundamental group of $\mathcal{M}\#\mathcal{N}$ is the free product $\mathbb{Z}/3\mathbb{Z} * \mathbb{Z}/3\mathbb{Z}$, which contains a free subgroup of rank 2 and is *not* virtually solvable. By Theorem 10.3, $\mathcal{M}\#\mathcal{N}$ admits no affine structure. This illustrates the profound interplay between algebra and geometry in affine topology.

Certain Seifert fibered 3-manifolds do not admit affine structures. This is characterized by:

Proposition 10.1 (Y. Carrière, F. Dal'bo, G. Meigniez, [CDM93]) *Let $\mathcal{M}$ be a Seifert fibered 3-manifold with vanishing first Betti number ($b_1(\mathcal{M}) = 0$). Then $\mathcal{M}$ admits no affine structure.*

10.2.5 Pre-Lie Structures

Affine structures lead to interesting algebraic properties on the tangent bundle/tangent sheaf. Let $\Gamma(\mathcal{T}(\mathcal{M}))$ denote the space of vector fields on a given manifold $\mathcal{M}$.

According to the previous chapters, a flat and torsionless connection satisfied the following. An affine connection ∇ is *torsion-free* (or torsionless) if

$$\nabla_X(Y) - \nabla_Y(X) - [X, Y] = 0. \tag{10.1}$$

The connection is called *flat* if

$$\nabla_X \nabla_Y(Z) - \nabla_Y \nabla_X(Z) - \nabla_{[X,Y]}(Z) = 0. \tag{10.2}$$

Such a connection defines a covariant differentiation for vector fields $X, Y \in \Gamma(\mathcal{T}(\mathcal{M}))$:

$$\nabla_X : \Gamma(\mathcal{T}(\mathcal{M})) \longrightarrow \Gamma(\mathcal{T}(\mathcal{M})), \quad \nabla_X(Y) \longmapsto \nabla_X(Y).$$

Remark 10.1 A pre-Lie algebra is an algebra satisfying the relation

$$a(bc) - b(ac) = (ab)c - (ba)c.$$

Pre-Lie algebras arise in deformation quantization, integrable systems, and renormalization. The connection ∇ induces such a structure on $\Gamma(\mathcal{T}(\mathcal{M}))$, linking differential geometry to algebra.

Exercise 10.1 Show that one can recover the structure of a pre-Lie algebra on the space of vector fields on $\mathcal{M}$, by putting $X \circ Y := \nabla_X(Y)$ for an affine flat, torsionless connection ∇.

Example 10.1 Let V be a finite-dimensional real Euclidean space endowed with a real inner product, and let C be a closed convex cone $C \subset V$ with its vertex at the origin. The polar of the cone is defined by

$$C^* = \{y \in V \mid \langle x, y \rangle > 0, \ \forall x \in C\}.$$

A symmetric cone satisfies $C^* = C$.

Such symmetric cones carry an affine flat structure. The tangent sheaf of this cone is equipped with the structure of a pre-Lie algebra, defined by the operation $X \circ Y := \nabla_X(Y)$, where X, Y are vector fields on C.

10.3 Flat Coordinates and Darboux–Egoroff Coordinates

Coordinates provide a foundational framework in geometry. This section examines two specialized types: **flat coordinates** and **Darboux–Egoroff coordinates**, both crucial in the theory of Frobenius manifolds.

Finding flat coordinates is essential when studying Gauss–Manin systems. These systems describe differential equations governing integrals of basic differential forms over vanishing cycles associated with singularities and are intimately linked to the theory of primitive forms. The Gauss–Manin connection provides a method to differentiate cohomology classes with respect to deformation parameters.

Flat Coordinates.

Consider a domain $\mathcal{U} \subset \mathbb{R}^n$ with local coordinates $x = (x^1, \dots, x^n)$ and a metric tensor g. Let $y = (y^1, \dots, y^n)$ be another coordinate system. The metric transforms as:

$$\tilde{g}_{ij}(y) = \sum_{k=1}^{n} \sum_{l=1}^{n} g_{kl}(x) \frac{\partial x^k}{\partial y^i} \frac{\partial x^l}{\partial y^j}.$$

The coordinates y are called **flat** if the components $\tilde{g}_{ij}(y)$ are *constant* on $\mathcal{U}$.

Kernel of the Metric.

At a point $x \in \mathcal{U}$, let $T_x\mathcal{M}$ denote the tangent space. The kernel of the metric g at x is defined as:

$$\ker(g|_x) := \{v \in T_x\mathcal{M} \mid g_x(v, w) = 0 \text{ for all } w \in T_x\mathcal{M}\}.$$

Christoffel Symbols.

The Levi-Civita connection associated with g is characterized by its Christoffel symbols. The symbols of the first kind are:

$$\Gamma_{ijk} := \frac{1}{2}\left(\frac{\partial g_{jk}}{\partial x^i} + \frac{\partial g_{ik}}{\partial x^j} - \frac{\partial g_{ij}}{\partial x^k}\right),$$

and the symbols of the second kind (connection coefficients) are:

$$\Gamma^k_{ij} := \frac{1}{2}\sum_{m=1}^{n} g^{km}\left(\frac{\partial g_{jm}}{\partial x^i} + \frac{\partial g_{im}}{\partial x^j} - \frac{\partial g_{ij}}{\partial x^m}\right),$$

where (g^{km}) is the inverse metric tensor.

Darboux–Egoroff Coordinates.

Darboux–Egoroff coordinates are a specific type of flat coordinate system $\mathbf{t} = (t^1, \ldots, t^n)$ satisfying the additional **Egoroff condition**:

$$\sum_{k=1}^{n} \frac{\partial \eta_k(\mathbf{t})}{\partial t^i} = 0 \quad \text{for all } i,$$

where $\boldsymbol{\eta} = \mathrm{diag}(\eta_1(\mathbf{t}), \ldots, \eta_n(\mathbf{t}))$ is the diagonalized metric tensor in these coordinates $(g_{ij} = \delta_{ij}\eta_i)$. These coordinates play a central role in the structure theory of Frobenius manifolds, particularly in defining the prepotential and associativity equations. Then, the following theorem holds.

Theorem 10.4 *Let g be a symmetric $(0, 2)$-tensor field on $\mathcal{U} \subset \mathbb{R}^n$ with constant rank. The following are equivalent for a symmetric connection ∇ $(\Gamma^k_{ij} = \Gamma^k_{ji})$:*

1. *The compatibility condition holds:*

$$\sum_{l=1}^{n}\left(\Gamma^l_{ki} g_{jl} + \Gamma^l_{kj} g_{il}\right) = \frac{\partial g_{ij}}{\partial x^k} \quad \forall i, j, k. \tag{10.3}$$

2. *For all $v \in \ker(g_x)$ and all i, k:*

$$\sum_{l=1}^{n} \Gamma^l_{ki} v^l = 0. \tag{10.4}$$

Moreover, if these conditions hold, there exists a smooth symmetric connection compatible with g.

Proof **Necessity** $(1 \Rightarrow 2)$**:** For $v \in \ker(g_x)$, differentiate $g(v, \cdot) = 0$:

$$0 = \frac{\partial}{\partial x^k}\left(\sum_m g_{im} v^m\right) = \sum_m \frac{\partial g_{im}}{\partial x^k} v^m + \sum_m g_{im} \frac{\partial v^m}{\partial x^k}.$$

Substitute (10.3):

$$0=\sum_m \left[\sum_l \left(\Gamma^l_{ki} g_{ml}+\Gamma^l_{km} g_{il}\right)\right] v^m+\sum_m g_{im}\frac{\partial v^m}{\partial x^k}.$$

Rearranging and using $\sum_m g_{ml} v^m=0$:

$$0=\sum_l \Gamma^l_{ki} \underbrace{\left(\sum_m g_{ml} v^m\right)}_{=0}+\sum_{m,l} \Gamma^l_{km} g_{il} v^m+\sum_m g_{im}\frac{\partial v^m}{\partial x^k}.$$

The last two terms represent $g(\nabla_k v, \cdot)$, proving (10.4) for the parallel kernel.

Sufficiency ($2 \Rightarrow 1$)**:** Fix i, k. The system (10.3) for all j is:

$$\sum_{l=1}^n \Gamma^l_{ki} g_{jl}=\underbrace{\frac{\partial g_{ij}}{\partial x^k}-\sum_{l=1}^n \Gamma^l_{kj} g_{il}}_{b_j}.$$

This is linear in $\mathbf{v}=(\Gamma^1_{ki}, \ldots, \Gamma^n_{ki})$. By the **Fredholm alternative**[2], solutions exist if and only if for all $a \in \ker(g_x)$:

$$\sum_j a^j b_j=0.$$

Condition (10.4) ensures this holds, as $a \in \ker(g_x)$ implies:

$$\sum_j a^j\left(\frac{\partial g_{ij}}{\partial x^k}-\sum_l \Gamma^l_{kj} g_{il}\right)=\sum_j a^j \frac{\partial g_{ij}}{\partial x^k}-g(\nabla_k a, \cdot)=0.$$

Smoothness follows from the constant rank condition. □

Theorem 10.5 *Let g be a non-degenerate metric. Flat coordinates for g exist if and only if there exists a symmetric connection* ∇ *satisfying:*

- ***Metric Compatibility:***

$$\nabla_k g_{ij}=\frac{\partial g_{ij}}{\partial x^k}-\sum_{l=1}^n\left(\Gamma^l_{ki} g_{lj}+\Gamma^l_{kj} g_{il}\right)=0.$$

[2] The Fredholm Alternative states that for a matrix equation $A\mathbf{x}=\mathbf{b}$, solutions exist if and only if $\langle\mathbf{a}, \mathbf{b}\rangle=0$ for all $\mathbf{a}$ in $\ker(A^\top)$. In the proof, for fixed i, k, the system $A\mathbf{v}=\mathbf{b}$ is defined by: $A_{jl}=g_{jl}$ (coefficient matrix); $\mathbf{v}=(\Gamma^1_{ki}, \ldots, \Gamma^n_{ki})$ (unknowns); $b_j=\partial_k g_{ij}-\sum_l \Gamma^l_{kj} g_{il}$ (right hand side). The adjoint condition requires $\sum_j a^j g_{jl}=0$ ($a \in \ker(g_x)$) implies $\sum_j a^j b_j=0$.

- ***Vanishing Curvature:***

$$R_{ijk}{}^{m} = \frac{\partial \Gamma^m_{jk}}{\partial x^i} - \frac{\partial \Gamma^m_{ik}}{\partial x^j} + \sum_{l=1}^{n} \left(\Gamma^m_{il}\Gamma^l_{jk} - \Gamma^m_{jl}\Gamma^l_{ik} \right) = 0.$$

Proof (⇒) Assume flat coordinates $\mathbf{y} = (y^1, \ldots, y^n)$ exist where $g_{ij} = c_{ij}$ (constant). Then:

- The Levi-Civita connection $\Gamma^k_{ij} = 0$ in $\mathbf{y}$-coordinates satisfies $\Gamma^k_{ij} = \Gamma^k_{ji}$ (symmetry)
- Metric compatibility holds trivially: $\nabla_k g_{ij} = \partial_k c_{ij} - \sum_l (0 \cdot g_{lj} + 0 \cdot g_{il}) = 0$
- Curvature vanishes: $R_{ijk}{}^{m} = \partial_i \Gamma^m_{jk} - \partial_j \Gamma^m_{ik} + \sum_l (0 \cdot \Gamma^l_{jk} - 0 \cdot \Gamma^l_{ik}) = 0$

(⇐) Assume there exists symmetric $\Gamma^k_{ij} = \Gamma^k_{ji}$ satisfying:

$$\begin{aligned} \nabla_k g_{ij} &= 0 \quad \text{(compatibility)} \\ R_{ijk}{}^{m} &= 0 \quad \text{(flatness)} \end{aligned}$$

Since ∇ is torsion-free ($\Gamma^k_{ij} = \Gamma^k_{ji}$) and flat, by the fundamental theorem of affine geometry:

1. There exist local coordinates $\mathbf{t} = (t^1, \ldots, t^n)$ where $\Gamma^k_{ij}(\mathbf{t}) = 0$ (affine coordinates).
2. In these coordinates, metric compatibility becomes:

$$\nabla_k g_{ij} = \frac{\partial g_{ij}}{\partial t^k} - \sum_l (0 \cdot g_{lj} + 0 \cdot g_{il}) = \frac{\partial g_{ij}}{\partial t^k} = 0.$$

3. Thus g_{ij} are constant in $\mathbf{t}$-coordinates.

Hence $\mathbf{t}$ are flat coordinates. Non-degeneracy of g ensures the coordinate transformation is well-defined. □

10.4 Pre-Frobenius Manifolds

Beyond their intrinsic appeal, pre–Frobenius manifolds stand at the confluence of several geometric traditions: their affine structure and compatible metric recall the special Kähler geometry of Calabi–Yau moduli, while the Monge–Ampère condition resonates with the theory of convex potentials in optimal transport and the affine differential geometry of hypersurfaces.

10.4.1 Essential Preliminary Ingredients

A pre-Frobenius manifold requires three essential components:

- An **affine flat structure**: an atlas where **flat coordinates** $\mathbf{x} = (x^1, \ldots, x^n)$ have affine-linear transition functions ($\phi_{ij}(\mathbf{x}) = A_{ij}\mathbf{x} + \mathbf{b}_{ij}$) and the associated connection ∇ satisfies $\nabla dx^k = 0$ (i.e., $\Gamma^k_{ij} = 0$).
- A **metric** g: a symmetric, non-degenerate (0, 2)-tensor that is **constant in flat coordinates**:

$$g = \sum_{i,j} g_{ij} dx^i \otimes dx^j \quad \text{with} \quad \partial_k g_{ij} = 0.$$

- A **symmetric rank-3 tensor** $A \in \Gamma(S^3\mathcal{T}^*(\mathcal{M}))$ that is **constant in flat coordinates**:

$$A = \sum_{i,j,k} A_{ijk} dx^i \otimes dx^j \otimes dx^k \quad \text{with} \quad \partial_\ell A_{ijk} = 0.$$

10.4.2 Multiplication Operation

In **flat coordinates**, the metric g and tensor A induce a commutative multiplication $\circ$ on the tangent sheaf via:

$$g(X \circ Y, Z) = A(X, Y, Z), \quad \forall X, Y, Z \in \mathcal{T}(\mathcal{M})$$

with local coordinate expression:

$$(X \circ Y)^k = \sum_\ell g^{k\ell} A_{ij\ell} X^i Y^j$$

where $g^{k\ell}$ is the inverse metric (also constant) and X, Y, Z are **flat** vector fields. The multiplication enjoys two fundamental properties: it is *commutative*,

$$X \circ Y = Y \circ X \quad \text{(a direct consequence of } A_{ijk} = A_{jik}),$$

and it is *g–symmetric*, in the sense that

$$g(X \circ Y, Z) = g(X, Y \circ Z) \quad \text{(as follows from } A_{ijk} = A_{ikj}).$$

10.4.3 Compatibility Relations

A compatibility relation between the rank-3 tensor A, the rank-2 tensor g, and the multiplication operation $\circ$ is given by:

$$\underbrace{A(X, Y, Z) =}_{\text{rank-3 tensor}} \underbrace{g(X \circ Y, Z)}_{\text{metric-multiplication pairing}} = \underbrace{g(X, Y \circ Z)}_{\text{associativity constraint}} .$$

This ensures the affine connection ∇ satisfies:

$$(\nabla_X g)(Y, Z) = 0 \quad \text{and} \quad \nabla A = 0.$$

This invariance of the metric with respect to multiplication ensures that the structure is well-defined.

Definition 10.2 A **pre-Frobenius manifold** is a quadruple $(\mathcal{M}, \nabla, g, A)$ satisfying the above-mentioned properties, that is:

- ∇ is a flat, torsion-free affine connection.
- g is a ∇-parallel metric.[3]
- A is a ∇-parallel symmetric 3-tensor.[4]
- The multiplication $\circ$ induced by g and A is commutative.

10.4.4 Potential Structure and Monge–Ampère Link

A critical refinement occurs when A derives from a potential. Namely, this axiom requires the existence of a family of local potentials Φ, such that:

$$g(X \circ Y, Z) = g(X, Y \circ Z) = (XYZ)\Phi.$$

This axiom is particularly important regarding the relations to the Monge–Ampère equations. We refer to [LV03, Vil03] for more information on the topic of Monge–Ampère domains.

[3] A metric tensor g (a symmetric, non-degenerate $(0, 2)$-tensor) is ∇-parallel if its covariant derivative vanishes identically: $\nabla g = 0$. In coordinates, this means that for all i, j, k:

$$\nabla_k g_{ij} := \frac{\partial g_{ij}}{\partial x^k} - \sum_{l=1}^{n} \left(\Gamma^l_{ki}\, g_{lj} + \Gamma^l_{kj}\, g_{il} \right) = 0.$$

Geometrically, g is invariant under parallel transport along any curve and the connection ∇ is compatible with g (for example, ∇ is the Levi-Civita connection associated to g).

[4] A symmetric 3-tensor A (i.e., $A_{ijk} = A_{jik} = A_{ikj} = \cdots$) is ∇-*parallel* if $\nabla A = 0$..

Definition 10.3 A pre-Frobenius manifold is **potential-type** if locally there exists $\Phi \in C^{\infty}(\mathcal{M})$ such that:

$$A(X, Y, Z) = \nabla_X \nabla_Y \nabla_Z \Phi = (XYZ\Phi)$$

Equivalently: $g(X \circ Y, Z) = (XYZ\Phi)$.

This potential-theoretic framework places us in the realm of elliptic Monge–Ampère equations, whose geometry is governed by the interplay between convexity and fully non-linear ellipticity. Consider a strictly convex domain $\mathscr{D} \subset \mathbb{R}^n$ with boundary data $\tilde{g} \in C(\partial\mathscr{D})$. The Monge–Ampère problem looks for $\Phi \in C^{\infty}(\mathscr{D})$ satisfying:

$$\begin{cases} \det(\text{Hess}\,\Phi) = f > 0 \quad \text{in } \mathscr{D}, \\ \Phi|_{\partial\mathscr{D}} = \tilde{g}. \end{cases} \tag{10.5}$$

From the analytic side, one encounters the delicate balance between the determinant constraint in (10.5) and the prescribed boundary data; from the geometric side, such solutions determine a canonical affine structure on $\mathscr{D}$. In order to work concretely within this setting, we make the following definition.

Definition 10.4 An **elliptic Monge–Ampère domain** is a pair $(\mathscr{D}, \Phi)$ where:

- $\mathscr{D} \subset \mathbb{R}^n$ is strictly convex.
- Φ is a strictly convex solution to (10.5).

Such domains inherit a rich affine and metric geometry whose local algebraic structure satisfies the axioms of a pre-Frobenius manifold.

This is the theorem below.

Theorem 10.6 *Every elliptic Monge–Ampère domain $(\mathscr{D}, \Phi)$ naturally carries a potential-type pre-Frobenius structure with ∇ a standard flat connection on $\mathbb{R}^n$, a Hessian metric $g_{ij} = \partial_i \partial_j \Phi$ and a cubic form $A_{ijk} = \partial_i \partial_j \partial_k \Phi$.*

Conversely, any simply connected potential-type pre-Frobenius manifold $(\mathcal{M}, \nabla, g, A)$ with:

- *a positive definite Hessian metric g;*
- *∇-parallel g and A ($\nabla g = 0$, $\nabla A = 0$);*
- *$A = \nabla^3 \Phi$ for local potential Φ;*

arises from such a domain. The correspondence is mediated by the elliptic Monge–Ampère equation $\det(\text{Hess}\,\Phi) > 0$.

Exercise 10.2 Prove Theorem 10.6 by establishing:

1. Forward Direction: Verify flatness of ∇, ∇-parallelism of g and A, and compatibility $A(X, Y, Z) = g(X \circ Y, Z)$.
2. Converse: Use ∇-parallelism to show g is Hessian, then solve $\det(\text{Hess}\,\Phi) = e^U$ for some ∇-parallel function U.

> While elliptic Monge–Ampère domains naturally give rise to pre-Frobenius manifold structures through their affine and metric geometry, elevating these to full Frobenius manifolds necessitates the fulfillment of a crucial extra condition: the **associativity** of the induced multiplication.

10.5 Frobenius Manifolds

10.5.1 Associative Pre-Frobenius Manifolds

A pre-Frobenius manifold is **associative** if its multiplication satisfies:

$$(X \circ Y) \circ Z = X \circ (Y \circ Z), \quad \forall X, Y, Z \in \mathcal{T}(\mathcal{M}).$$

This induces a rigid constraint on the cubic tensor A:

$$A(X \circ Y, Z, W) = A(X, Y \circ Z, W) = A(X, Y, Z \circ W).$$

When combined with potentiality ($A = \nabla^3\Phi$), associativity leads to Frobenius manifolds.

> A **Frobenius manifold** is therefore a potential-type pre-Frobenius manifold satisfying the associativity axiom.

From a more analytic perspective, the associativity condition on the product translates into a highly nontrivial system of nonlinear partial differential equations–the Witten–Dijkgraaf–Verlinde–Verlinde (WDVV) equations–thus bridging the geometric framework with integrable structures and topological field theory.

10.5.2 Witten–Dijkgraaf–Verlinde–Verlinde Equation

For a Frobenius manifold endowed with a potential function Φ, the associativity of the product imposes a rich and highly structured system of partial differential equations known as the WDVV equations. Let $\{\partial_a\}$ denote a ∇-flat local frame on the tangent bundle. The multiplication is expressed as

$$\partial_a \circ \partial_b = \sum_c c_{ab}^c \partial_c, \quad \text{where} \quad c_{ab}^c = \sum_d g^{cd}\, \partial_a \partial_b \partial_d \Phi,$$

with g^{cd} the inverse of the metric $g_{cd} = \partial_c \partial_d \Phi$. The associativity condition

$$(\partial_a \circ \partial_b) \circ \partial_c = \partial_a \circ (\partial_b \circ \partial_c)$$

then translates into the celebrated nonlinear system

$$\sum_{e,f} (\partial_a \partial_b \partial_e \Phi)\, g^{ef}\, (\partial_f \partial_c \partial_d \Phi) = \sum_{e,f} (\partial_b \partial_c \partial_e \Phi)\, g^{ef}\, (\partial_f \partial_a \partial_d \Phi), \tag{10.6}$$

holding for all indices a, b, c, d. This system, known as the Witten–Dijkgraaf–Verlinde–Verlinde (WDVV) equations, comprises a set of third-order nonlinear PDEs quadratic in the third derivatives of Φ. Solutions to (10.6) are in one-to-one correspondence with Frobenius manifold structures.

10.5.3 Geometrization of Frobenius Structures

To ground our discussion in a more rigorous framework, we now introduce a more explicit definition of Frobenius manifolds, which can be viewed as a geometric incarnation of what has been already discussed using the notion of Frobenius algebras. Recall that a Frobenius algebra $(\mathcal{A}, \circ)$ over a field $\mathbb{K}$ is a finite-dimensional, commutative, associative, unital algebra endowed with a symmetric bilinear form $\langle \cdot, \cdot \rangle$ satisfying the crucial invariance property

$$\langle x \circ y, z \rangle = \langle x, y \circ z \rangle, \quad \forall x, y, z \in \mathcal{A}.$$

This compatibility condition tightly couples the algebraic multiplication and the bilinear form, encoding a rich duality at the heart of the structure.

Definition 10.5 A **Frobenius manifold** is a quadruple $(\mathcal{M}, \nabla, g, \circ)$ where $\mathcal{M}$ is assumed to be an affine manifold and where:

- ∇ is a flat, torsion-free connection on the tangent bundle $\mathcal{T}(\mathcal{M})$.
- g is a ∇-parallel metric tensor.

- $\circ$ is a commutative, associative multiplication on $\mathcal{T}(\mathcal{M})$.
- g is invariant under $\circ$, that is,

$$g(X \circ Y, Z) = g(X, Y \circ Z) \quad \text{for all flat } X, Y, Z \in \Gamma(\mathcal{T}(\mathcal{M}));$$

- there exists a ∇-flat identity vector field $e \in \Gamma(\mathcal{T}(\mathcal{M}))$ satisfying

$$e \circ X = X, \quad \forall X \in \Gamma(\mathcal{T}(\mathcal{M}));$$

- locally, the structure is governed by a potential function Φ such that

$$g_{ij} = \partial_i \partial_j \Phi, \quad A_{ijk} = \partial_i \partial_j \partial_k \Phi,$$

  where A denotes the structure constants of the multiplication.

Equivalent Characterization via Tangent-Algebra Compatibility

A *Frobenius algebra* is a finite-dimensional unital associative algebra $(\mathcal{A}, \circ, 1)$ over a field $\mathbb{K}$, together with a linear functional $\epsilon : \mathcal{A} \to \mathbb{K}$ (the *Frobenius form*) such that there exists an **associative non-degenerate bilinear form** $\beta : \mathcal{A} \times \mathcal{A} \to \mathbb{K}$ such that

$$\beta(a \circ b, c) = \beta(a, b \circ c), \quad \forall a, b, c \in \mathcal{A}.$$

To summarize, we can say that:

> A manifold $\mathcal{M}$ admitting a Frobenius structure is such that at each point $p \in \mathcal{M}$ the tangent space $\mathcal{T}_p(\mathcal{M})$ is endowed with a Frobenius algebra structure $(\mathcal{T}_p(\mathcal{M}), \circ_p, \langle \cdot, \cdot \rangle_p)$ that varies smoothly with the point $p \in \mathcal{M}$.

Furthermore, there exists an **Euler vector field** E satisfying the conditions:

$$\nabla(\nabla E) = 0 \quad \text{(affineness)}, \tag{10.7}$$

$$\mathcal{L}_E(\circ) = \mu \circ \quad \text{(quasi-homogeneity)}, \quad \mu \in \mathbb{R}, \tag{10.8}$$

where the Lie derivative $\mathcal{L}_E(\circ)$ measures how the multiplication operation $\circ$ changes along the flow of the vector field E. Explicitly, for vector fields X, Y:

$$(\mathcal{L}_E \circ)(X, Y) := \mathcal{L}_E(X \circ Y) - (\mathcal{L}_E X) \circ Y - X \circ (\mathcal{L}_E Y)$$

This evaluates the infinitesimal deformation of $\circ$ under E.

The notion of affineness for the Euler vector field E in the setting of Frobenius manifolds admits several equivalent characterizations. Concretely, the following conditions are equivalent:

1. E is affine in the strongest sense, satisfying

$$\nabla(\nabla E) = 0,$$

that is, the second covariant derivative vanishes identically.

2. For arbitrary vector fields Y, Z, the covariant derivatives commute with ∇E:

$$\nabla_Y(\nabla_Z E) = \nabla_{\nabla_Y Z} E,$$

expressing compatibility of ∇E with the connection ∇.

3. In any ∇-flat coordinate system $\{x^m\}$, the components of E are affine functions:

$$E^m = \sum_j a^m_j x^j + b^m, \quad a^m_j, b^m \in \mathbb{R},$$

exhibiting linearity plus translation in these distinguished coordinates.

4. The flow ϕ_t generated by E preserves the affine structure, i.e.,

$$\phi_t^* \nabla = \nabla,$$

so that parallel transport is invariant under the Euler flow.

This equivalence reveals the profound interplay between the affine geometry encoded by ∇ and the scaling symmetries driven by the Euler field.

10.5.4 Structure Connections of Pre-Frobenius Manifolds

Let $(\mathcal{M}, \nabla_0, g, A)$ be a pre-Frobenius manifold.

- The **base connection** ∇_0 has the property that its flat sections s (those satisfying $\nabla_0 s = 0$) are constant in affine coordinates.
- The **structure connection** is a pencil of connections parameterized by $\lambda \in \mathbb{C}$:

$$\nabla^\lambda_X Y := \nabla_{0,X} Y + \lambda\, X \circ Y. \tag{10.9}$$

This deforms ∇_0 by the multiplication $\circ$ scaled by λ.

Theorem 10.7 (Manin) *Let $(\mathcal{M}, g, A, \nabla_\lambda)$ be a pre-Frobenius manifold equipped with a structure connection ∇_λ. $(\mathcal{M}, g, A)$ is a Frobenius manifold if and only if the pencil $\{\nabla_\lambda\}$ is flat.*

Exercise 10.3 Prove the Theorem 10.7.

10.6 Emergence of Frobenius Structures on Statistical Manifolds

This section provides a brief overview of how statistical manifolds naturally induce Frobenius algebra structures on their tangent spaces. The full mathematical development will follow in subsequent chapters.

10.6.1 The Fundamental Tensors

Let $(\mathscr{S}, g)$ be a statistical manifold of exponential type with:

- Fisher metric $g_{ij} = \mathbb{E}_\theta[\partial_i \ell_\theta \partial_j \ell_\theta]$ where we recall that $\ell_\theta = \ln \rho_\theta$ and ρ_θ is a probability distribution (see previous Chapters). The $\partial_i \ell_\theta$ are components of the gradient of ℓ_θ, the so-called *score vector*.[5]
- We have the skewness tensor given by $\mathbf{T}_{ijk} = \mathbb{E}_\theta[\partial_i \ell_\theta \partial_j \ell_\theta \partial_k \ell_\theta]$.

Define the *Amari–Chentsov mixed tensor* $\bar{\mathbf{T}}$ as the $(1, 2)$-tensor obtained by raising one index of $\mathbf{T}$:

$$\bar{\mathbf{T}}_{ij}^{k} = \sum_m g^{km} \mathbf{T}_{ijm}$$

where g^{km} is the inverse Fisher metric. This tensor encodes the cubic interactions in the tangent space.

10.6.2 The Hidden Multiplication

At each point $P_\theta \in \mathscr{S}$, the tensor $\bar{\mathbf{T}}$ induces a commutative multiplication $\circ$ on the tangent space $\mathcal{T}_{P_\theta}(\mathscr{S})$:

[5] Let $\Theta \subseteq \mathbb{R}^k$ be a parameter space, X a random vector with distribution ρ_θ, and $\ell_\theta(X)$ the log-likelihood function. The *score vector* $\mathbf{U}(\theta)$ is the gradient of ℓ with respect to θ:

$$\mathbf{U}(\theta) = \left(\frac{\partial \ell}{\partial \theta_1}, \ldots, \frac{\partial \ell}{\partial \theta_k}\right)^{\top}.$$

It is important in relation to the maximum likelihood estimator.

$$\circ : \mathcal{T}_{P_\theta}(\mathcal{S}) \times \mathcal{T}_{P_\theta}(\mathcal{S}) \to \mathcal{T}_{P_\theta}(\mathcal{S}), \quad (u, v) \mapsto u \circ v$$

defined by:

$$(u \circ v)^k = \sum_{i,j} \bar{T}^k_{ij} u^i v^j$$

Theorem 10.8 (Emergent Frobenius Structure) *For $u, v, w \in \mathcal{T}_{P_\theta}(\mathcal{S})$, the multiplication $\circ$ satisfies:*

1. ***Commutativity**: $u \circ v = v \circ u$*
2. ***Metric Compatibility**: $g(u \circ v, w) = g(u, v \circ w)$*
3. ***Tangent Space Algebra**: $(\mathcal{T}_{P_\theta}(\mathcal{S}), \circ, g)$ forms a commutative Frobenius algebra*

We proceed to a statistical interpretation of the statement before providing a proof.

In terms of score vectors:

$$g(u \circ v, w) = \mathbb{E}_\theta[(u^i \partial_i \ell_\theta)(v^j \partial_j \ell_\theta)(w^k \partial_k \ell_\theta)]$$

This shows the multiplication encodes third-order statistical interactions. The Frobenius condition

$$g(u \circ v, w) = g(u, v \circ w)$$

reflects the duality between different decompositions of cubic statistical interactions.

Remark 10.2 For exponential families, this structure is naturally flat with ∇_0 being the exponential connection. The associativity of $\circ$ corresponds to the vanishing of the fourth cumulant tensor, which will be explored in Chap. 7.

Proof 1. **Commutativity**: $u \circ v = v \circ u$. The multiplication is defined component-wise as:

$$(u \circ v)^k = \sum_{i,j} \bar{T}^k_{ij} u^i v^j, \quad \text{where } \bar{\mathbf{T}}^k_{ij} = \sum_m g^{km} \mathbf{T}_{ijm}.$$

The skewness tensor $\mathbf{T}_{ijm}$ is symmetric in all indices ($\mathbf{T}_{ijm} = \mathbf{T}_{jim} = \mathbf{T}_{imj} = \cdots$) because

$$\mathbf{T}_{ijm} = \mathbb{E}_\theta[\partial_i \ell_\theta \cdot \partial_j \ell_\theta \cdot \partial_m \ell_\theta],$$

and partial derivatives commute ($\partial_i \partial_j \ell_\theta = \partial_j \partial_i \ell_\theta$). Thus, $\bar{\mathbf{T}}^k_{ij} = \bar{\mathbf{T}}^k_{ji}$.
This implies:

$$(v \circ u)^k = \sum_{i,j} \bar{\mathbf{T}}^k_{ij} v^i u^j = \sum_{i,j} \bar{\mathbf{T}}^k_{ji} u^j v^i = \sum_{i,j} \bar{\mathbf{T}}^k_{ij} u^i v^j = (u \circ v)^k.$$

and therefore commutativity follows.

2. We discuss now the **metric compatibility**: $g(u \circ v, w) = g(u, v \circ w)$. The Fisher metric g gives:

$$g(u \circ v, w) = \sum_{k,\ell} g_{k\ell}(u \circ v)^k w^\ell = \sum_{i,j,k,\ell} g_{k\ell} \bar{\mathbf{T}}_{ij}^k u^i v^j w^\ell.$$

Substitute $\bar{\mathbf{T}}_{ij}^k = \sum_m g^{km} \mathbf{T}_{ijm}$:

$$g(u \circ v, w) = \sum_{i,j,k,\ell,m} g_{k\ell} g^{km} \mathbf{T}_{ijm} u^i v^j w^\ell.$$

Since $\sum_k g_{k\ell} g^{km} = \delta_\ell^m$ (inverse metric property):

$$g(u \circ v, w) = \sum_{i,j,\ell} \mathbf{T}_{ij\ell} u^i v^j w^\ell.$$

Similarly:

$$g(u, v \circ w) = \sum_{i,\ell} g_{i\ell} u^i (v \circ w)^\ell = \sum_{i,j,\ell,m} g_{i\ell} \bar{\mathbf{T}}_{jm}^\ell u^i v^j w^m.$$

Substitute $\bar{\mathbf{T}}_{jm}^\ell = \sum_n g^{\ell n} \mathbf{T}_{jmn}$:

$$g(u, v \circ w) = \sum_{i,j,\ell,m,n} g_{i\ell} g^{\ell n} \mathbf{T}_{jmn} u^i v^j w^m = \sum_{i,j,m} \mathbf{T}_{jmi} u^i v^j w^m.$$

By symmetry of $\mathbf{T}$ (that is: $\mathbf{T}_{jmi} = \mathbf{T}_{ijm}$):

$$g(u, v \circ w) = \sum_{i,j,m} \mathbf{T}_{ijm} u^i v^j w^m = g(u \circ v, w).$$

Metric compatibility holds.

3. **Commutative Frobenius Algebra**
The triple $(\mathcal{T}_{P_\theta}(\mathcal{S}), \circ, g)$ forms a commutative Frobenius algebra because we have:

- A *Commutative Algebra*: $\circ$ is bilinear (tensor definition), commutative (proven), and associative (shown below).
- *Frobenius Property*: g is non-degenerate (Fisher metric) and invariant (proven).

Associativity: For exponential families, the fourth cumulant vanishes, ensuring:

$$(u \circ v) \circ w = u \circ (v \circ w).$$

This follows from the identity:

$$\mathbb{E}_\theta[(\partial_a \ell_\theta)(\partial_b \ell_\theta)(\partial_c \ell_\theta)(\partial_d \ell_\theta)] = \mathbf{T}_{abc}\mathbf{T}_{def}g^{ef} + \text{lower terms},$$

which reduces to the WDVV equation when the fourth cumulant is zero.
Thus, $(\mathcal{T}_{P_\theta}(\mathscr{S}), \circ, g)$ is a commutative Frobenius algebra. □

10.6.3 Summary of Key Structures

We sumarize the key elements existing for pre-Frobenius manifolds, in the example of statistical manifolds. This is as follows:

- An affine flat structure. This structure exists for statistical manifolds of exponential type.
- A triple $(\mathscr{S}, g, \mathbf{T})$, where $\mathscr{S}$ is a finite-dimensional statistical manifold:

- g is a Riemannian (Fisher) metric.
- $\mathbf{T}$ is the rank-3 symmetric Amari–Chentsov tensor.
- A multiplication defined by:
$$u \circ v = \bar{\mathbf{T}}(u, v),$$
where locally,
$$\bar{\mathbf{T}}_{ij}^k = \sum g^{km}\mathbf{T}_{ijm},$$
- Metric invariance:
$$g(u, v \circ w) = g(u \circ v, w), \quad \text{for flat vector fields} \quad u, v, w.$$
- The flatness condition, which requires that the affine space of connections ∇_λ is flat.

10.7 Semisimple Frobenius Manifolds

Semisimple Frobenius manifolds play a fundamental role in the theory of integrable systems, singularity theory, and quantum cohomology. Their structure is deeply connected to Saito's theory of primitive forms and configuration spaces of Coxeter systems.

10.7.1 Definition and Characterizations

Let $(\mathcal{M}, \nabla, g, \circ)$ be a Frobenius manifold of dimension n.

Definition 10.6 $\mathcal{M}$ is **semisimple** if at each point $p \in \mathcal{M}$, the Frobenius algebra $(\mathcal{T}_p(\mathcal{M}), \circ_p, g_p)$ is semisimple (i.e., contains no nilpotent elements). Equivalently, there exists:

1. A local isomorphism of $\mathcal{O}_{\mathcal{M}}$-algebras:

$$(\mathcal{T}(\mathcal{M}), \circ) \xrightarrow{\sim} \mathcal{O}_{\mathcal{M}}^{\oplus n}$$

   where the right side has componentwise multiplication.
2. Let $(\mathbf{e}_1, \mathbf{e}_2, \ldots, \mathbf{e}_n)$ be a local basis of $\mathcal{T}(\mathcal{M})$. In this local basis, the multiplication takes the form:

$$\left(\sum f_i \mathbf{e}_i\right) \circ \left(\sum g_j \mathbf{e}_j\right) = \sum f_i g_i \mathbf{e}_i.$$

   In the simplest case, this reduces to

$$\mathbf{e}_i \circ \mathbf{e}_j = \delta_{ij} \mathbf{e}_i.$$

   Thus, the basis $(\mathbf{e}_i)$ provides a well-defined (up to renumbering) family of idempotents. If $\mathcal{M}$ is semisimple, there exists an unramified covering of $\mathcal{M}$ (of degree at most $n!$) on which the induced pre-Frobenius structure becomes a splitting structure.

10.7.2 Canonical Decomposition in Semisimple Case

Let $\{\mathbf{e}_i\}$ be a local basis of idempotents for $(\mathcal{T}(\mathcal{M}), \circ)$ satisfying $\mathbf{e}_i \circ \mathbf{e}_j = \delta_{ij}\mathbf{e}_i$, and let $\{\epsilon^i\}$ be the dual basis of 1-forms. The structure tensor A, encoding the Frobenius structure, satisfies:

$$A(X, Y, Z) = g(X \circ Y, Z) = g(X, Y \circ Z)$$

where g is the flat metric and $\circ$ is the associative multiplication. Applying this to idempotents yields:

$$g(\mathbf{e}_i, \mathbf{e}_j) = g(\mathbf{e}_i \circ \mathbf{e}_i, \mathbf{e}_j) = g(\mathbf{e}_i, \mathbf{e}_i \circ \mathbf{e}_j) = \delta_{ij} g(\mathbf{e}_i, \mathbf{e}_i).$$

Define $\eta_i := g(\mathbf{e}_i, \mathbf{e}_i) > 0$ (since g is positive definite). The tensor A then decomposes as:

$$A = \sum_{i=1}^{n} \eta_i (\epsilon^i \otimes \epsilon^i \otimes \epsilon^i).$$

The identity element $\mathbf{e} = \sum_{i=1}^{n} \mathbf{e}_i$ has the co-identity (a 1-form dual to the identity):

$$\mathcal{E} = \sum_{i=1}^{n} \eta_i \epsilon^i$$

which satisfies $\mathcal{E}(\mathbf{e}) = \sum \eta_i = \dim \mathcal{M}$.

As a consequence of the above discussion, we are able to reformulate our initial definition differently.

Definition 10.7 A *semisimple Frobenius structure* on a smooth manifold $\mathcal{M}$ consists of the following data:

1. a reduction of the structure group of the tangent sheaf $\mathcal{T}(\mathcal{M})$;
2. a flat metric g, diagonal in a distinguished basis $(\mathbf{e}_i)$ and its dual (ϵ^i);
3. a diagonal cubic tensor A, sharing the same coefficients as g.

Remark 10.3 It is interesting to note that while the conditions of potentiality and flatness imposed on g are of a non-trivial nature, the associativity of the product structure on $\mathcal{T}(\mathcal{M})$ follows automatically under these hypotheses.

10.7.3 Splitting Structure

We now turn to a more detailed exploration of the facts introduced above.

The semisimplicity condition implies the following.

Theorem 10.9 *For a generic*[6] *semisimple Frobenius manifold $\mathcal{M}$, there exists an unramified covering $\pi : \widetilde{\mathcal{M}} \to \mathcal{M}$ of degree smaller or equal to $n!$ such that*

- *the pullback Frobenius structure on $\widetilde{\mathcal{M}}$ is* ***split semisimple****:*

$$\pi^*(\mathcal{T}(\mathcal{M}), \circ) \cong \mathcal{O}_{\widetilde{\mathcal{M}}}^{\oplus n}.$$

- *Canonical coordinates (or Darboux–Egoroff coordinates) are special local coordinates $(u^1, \cdots, u^n)$ which exist globally on $\widetilde{\mathcal{M}}$.*
- *The Euler vector field takes diagonal form:*

$$E = \sum_{i=1}^{n} u^i \frac{\partial}{\partial u^i},$$

where $\mathbf{e}_i = \frac{\partial}{\partial u^i}$.

Proof Let $\mathcal{M}$ be a generic semisimple Frobenius manifold of dimension n (i.e., eigenvalues $\{u^i\}$ distinct everywhere). We explicitly construct the covering space $\widetilde{\mathcal{M}}$ as the **space of ordered idempotent bases**:

[6] This holds when $\{u^i\}$ is simple (all u^i distinct).

$$\widetilde{\mathcal{M}} = \left\{ \left(p, \{\mathbf{e}_1, \ldots, \mathbf{e}_n\}\right) \mid p \in \mathcal{M},\ \mathbf{e}_i \in \mathcal{T}_p(\mathcal{M}),\ \mathbf{e}_i \circ \mathbf{e}_j = \delta_{ij}\mathbf{e}_i \right\},$$

with projection $\pi : (p, \{\mathbf{e}_i\}) \mapsto p$. The symmetric group S_n acts freely on $\widetilde{\mathcal{M}}$ by permuting the basis:

$$\sigma \cdot (p, \{\mathbf{e}_1, \ldots, \mathbf{e}_n\}) = (p, \{\mathbf{e}_{\sigma(1)}, \ldots, \mathbf{e}_{\sigma(n)}\}), \quad \sigma \in S_n.$$

Since all u^i are distinct, the fiber $\pi^{-1}(p)$ has exactly $|S_n| = n!$ points, making π an unramified covering of degree $n!$.

Step 1: Split Semisimplicity
The pullback tangent bundle $\pi^*\mathcal{T}(\mathcal{M})$ admits a global frame $\{\widetilde{\mathbf{e}}_1, \ldots, \widetilde{\mathbf{e}}_n\}$ defined by:

$$\widetilde{\mathbf{e}}_i\left((p, \{\mathbf{e}_j\})\right) = \mathbf{e}_i.$$

By construction, $\widetilde{\mathbf{e}}_i \circ \widetilde{\mathbf{e}}_j = \delta_{ij}\widetilde{\mathbf{e}}_i$. This trivializes the algebra structure:

$$(\pi^*\mathcal{T}(\mathcal{M}), \circ) \cong \bigoplus_{i=1}^{n} \mathcal{O}_{\widetilde{\mathcal{M}}} \cdot \widetilde{\mathbf{e}}_i \cong \mathcal{O}_{\widetilde{\mathcal{M}}}^{\oplus n}.$$

Step 2: Global Canonical Coordinates
Define functions $u^i : \widetilde{\mathcal{M}} \to \mathbb{C}$ via the Euler field:

$$E \circ \widetilde{\mathbf{e}}_i = u^i \widetilde{\mathbf{e}}_i.$$

Since $\{u^i\}$ are distinct and $\widetilde{\mathbf{e}}_i$ commute ($[\widetilde{\mathbf{e}}_i, \widetilde{\mathbf{e}}_j] = 0$), the map $(u^1, \ldots, u^n) : \widetilde{\mathcal{M}} \to \mathbb{C}^n$ is étale (i.e. has no singular/ramifications points). Thus, (u^i) are global coordinates with:

$$\widetilde{\mathbf{e}}_i = \frac{\partial}{\partial u^i}.$$

Step 3: Diagonal Euler Field
In coordinates (u^i), the Euler field satisfies:

$$E \circ \frac{\partial}{\partial u^i} = u^i \frac{\partial}{\partial u^i}.$$

As $\{u^i\}$ are distinct, E is diagonalizable with:

$$E = \sum_{i=1}^{n} u^i \frac{\partial}{\partial u^i}.$$

Step 4: Flatness and Compatibility
The Levi-Civita connection ∇ satisfies $\nabla \mathbf{e} = 0$ for $\mathbf{e} = \sum_i \frac{\partial}{\partial u^i}$, and the metric is diagonal:

$$g\left(\frac{\partial}{\partial u^i}, \frac{\partial}{\partial u^j}\right) = \delta_{ij}\eta_i(u).$$

Thus, $\widetilde{\mathcal{M}}$ carries a split semisimple Frobenius structure with global canonical coordinates. □

10.7.4 Geometric Interpretation: Deformations and Isomonodromy

We now turn briefly to the geometric interpretation of semisimple Frobenius manifolds, whose idempotent decomposition reveals deep ties to the theory of isomonodromic deformations. Although this rich correspondence constitutes a vast subject in its own right–intertwining algebraic geometry, integrable systems, and the analytic theory of Painlevé equations–we confine ourselves here to a concise overview, leaving its full development beyond the scope of the present work.

1. Diagonalizability

The condition that operators $X \circ (\cdot)$[7] are *simultaneously diagonalizable* means that at each point $p \in \mathcal{M}$, the Frobenius algebra $(\mathcal{T}_p(\mathcal{M}), \circ)$ decomposes into orthogonal 1D subalgebras:

$$\mathcal{T}_p(\mathcal{M}) \cong \bigoplus_{i=1}^{n} \mathbb{C} \cdot \mathbf{e}_i.$$

As a **consequence**, the spectrum of $X \circ (\cdot)$ determines **canonical coordinates** $\{u^i\}$ via:

$$X \circ \mathbf{e}_i = u^i \mathbf{e}_i,$$

where u^i are eigenvalues of the operator.

2. Flat Identity

The condition $\nabla \mathbf{e} = 0$ (where $\mathbf{e} = \sum_i \mathbf{e}_i$) implies that the identity vector field $\mathbf{e}$ is covariantly constant under the Levi-Civita connection ∇ of g. In local coordinates this amounts to having

$$\nabla_k \mathbf{e}^j = \partial_k \mathbf{e}^j + \Gamma^j_{k\ell}\mathbf{e}^\ell = 0.$$

---

[7] For a fixed tangent vector $X \in \mathcal{T}_p(\mathcal{M})$, the map

$$X \circ (\cdot)\colon \mathcal{T}_p(\mathcal{M}) \longrightarrow \mathcal{T}_p(\mathcal{M}), \quad Y \longmapsto X \circ Y,$$

defines a linear endomorphism of the tangent space at p. This endomorphism is referred to as the *multiplication operator* by X.

As a **geometric consequence**, we find that $\mathbf{e}$ generates a flat direction in $\mathcal{M}$. The Euler vector field $E = \sum u^i \partial_{u^i}$ satisfies:

$$\mathcal{L}_E \mathbf{e} = 0.$$

3. Deformation Rigidity and Isomonodromy

In the semisimple theory of Frobenius manifolds, the algebraic decomposition into idempotents is not merely a formal device–it is the natural stage on which the geometry of deformations unfolds.

At a semisimple point of a Frobenius manifold $\mathcal{M}$, the tangent space $\mathcal{T}_p(\mathcal{M})$ admits a distinguished basis $\{\mathbf{e}_1, \ldots, \mathbf{e}_n\}$ consisting of *idempotent vector fields*. These fields span $\mathcal{T}_p(\mathcal{M})$ and enjoy the property that the metric g is diagonal in this basis.

The Euler vector field E acts diagonally with respect to the idempotent basis, and the $u^1, \ldots, u^n$ are given by the eigenvalues of the Euler operator:

$$E \circ \mathbf{e}_i = u_i \, \mathbf{e}_i .$$

In this setting, the *idempotent frame* $\{\mathbf{e}_i\}$ evolves according to the celebrated *Schlesinger equations*:

$$\frac{\partial \mathbf{e}_i}{\partial u^j} = \begin{cases} 0, & \text{if } i = j, \\ \dfrac{[\mathbf{e}_i, \mathbf{e}_j]}{u^i - u^j}, & \text{if } i \neq j. \end{cases}$$

These relations encode a remarkably rigid deformation pattern: each $\mathbf{e}_i$ remains stationary along its own coordinate direction, while its variation along u^j is entirely determined by the commutator with $\mathbf{e}_j$, modulated by the simple pole $(u^i - u^j)^{-1}$.

From the viewpoint of integrable systems, these equations arise as the compatibility conditions for the *isomonodromic deformation* of the Fuchsian system

$$\frac{\mathrm{d}\Psi}{\mathrm{d}z} = \left(\sum_{i=1}^{n} \frac{\mathbf{e}_i}{z - u^i} \right) \Psi,$$

where Ψ is a matrix-valued solution and $n = \dim(\mathcal{M})$. The Schlesinger flow preserves the monodromy data of this connection, so that the deformation of the pole positions u^i leaves the analytic continuation structure invariant.

This preservation of monodromy is not merely a curiosity: it places Frobenius manifold deformations squarely inside the landscape of classical integrability. Indeed, the Schlesinger equations are Hamiltonian with respect to the Dubrovin–Krichever Poisson structure, and their special solutions are expressed in terms of *Painlevé transcendents*–notably Painlevé VI in the four-pole case.

The interplay between Frobenius geometry and isomonodromy can thus be distilled into the following schematic:

Condition	Role in frobenius manifolds	Link to integrable systems
Diagonalizability	Semisimplicity $\Leftrightarrow$ Local splitting into 1D algebras	Spectral parameter in Lax pairs
Flat identity	Compatibility of $\circ$ with flat structure	Conservation laws
Isomonodromy equations	Deformation rigidity $\Leftrightarrow$ Monodromy preservation	Solutions as Painlevé transcendents

For a more complete description relating to integrable systems we refer to [D96, JM81, Sa07].

10.7.5 Axiomatic Definition of Semisimple Frobenius Structures

To conclude, we arrive at the definition in its final, distilled form, trusting that the preceding discussion has rendered it more natural.

Definition 10.8 A (non-generic) **semisimple Frobenius structure** on a manifold $\mathcal{M}$ of dimension n consists of:

1. A **commutative associative multiplication** $\circ$ on $\mathcal{T}(\mathcal{M})$ with unit $\mathbf{e} \in \Gamma(\mathcal{T}(\mathcal{M}))$ satisfying $\nabla \mathbf{e} = 0$ (flat identity).
2. A **flat metric** g compatible with $\circ$:
$$g(X \circ Y, Z) = g(X, Y \circ Z), \quad \forall X, Y, Z \in \mathcal{T}(\mathcal{M})$$
3. An **Euler vector field** E satisfying:
$$\mathcal{L}_E(\circ) = \mu \circ \quad \text{and} \quad \mathcal{L}_E g = (2-d)g$$
for some $\mu, d \in \mathbb{C}$.
4. **Semisimplicity**: At generic points, there exist local coordinates $(u^1, \dots, u^n)$ where:
$$\begin{aligned}
\mathbf{e}_i &:= \frac{\partial}{\partial u^i} \quad \text{(idempotents)}, \\
\mathbf{e}_i \circ \mathbf{e}_j &= \delta_{ij}\mathbf{e}_i, \\
g(\mathbf{e}_i, \mathbf{e}_j) &= \delta_{ij}\eta_i(u), \\
E &= \underbrace{\sum_{i=1}^{n} u_i\, e_i}_{\text{diagonal part}} \;+\; \underbrace{\sum_{\alpha=1}^{k} c_\alpha\, v_\alpha}_{\text{nilpotent part}},
\end{aligned}$$

where

- $c_\alpha \in \mathbb{C}$ are constants.
- v_α are nilpotent vector fields satisfying:

$$[e_i, v_\alpha] = 0, \quad v_\alpha \circ v_\beta = 0, \quad g(v_\alpha, \cdot) = 0.$$

The v_α span the kernel of the algebra at points where the multiplication degenerates.

The cubic tensor $A(X, Y, Z) := g(X \circ Y, Z)$ satisfies $A = \sum_{i=1}^n \eta_i (du^i)^{\otimes 3}$ in canonical coordinates.

Remark 10.4 The non-generic situation arises, for instance, at points of spectral degeneracy, where $u_i = u_j$ for some $i \neq j$–the so-called discriminant locus. A further manifestation occurs in the presence of non-trivial monodromy, whereby the eigenvalues undergo permutations or shifts upon analytic continuation along loops in $\mathcal{M}$.

10.7.6 Characterization of Semisimple Frobenius Structures

We now establish a fundamental characterization of semisimple Frobenius structures, highlighting the interplay between multiplication, metric, and compatibility conditions.

Theorem 10.10 *A semisimple pre-Frobenius manifold* $(\mathcal{M}, \nabla_0, g, \circ)$ *defines a Frobenius manifold if and only if:*

1. *The idempotent vector fields* $\{\mathbf{e}_i\}$ *commute:* $[\mathbf{e}_i, \mathbf{e}_j] = 0$ *for all* i, j. *This implies the existence of canonical coordinates* $\{u^i\}$ *where* $\mathbf{e}_i = \partial/\partial u^i$ *and* $\epsilon^i = du^i$.
2. *The metric coefficients* $\eta_i = g(\mathbf{e}_i, \mathbf{e}_i)$ *satisfy* $d\mathcal{E} = 0$ *for the co-identity* $\mathcal{E} = \sum_{i=1}^n \eta_i \epsilon^i$, *or equivalently, there exists a local function* η *such that* $\eta_i = \mathbf{e}_i(\eta)$.

Proof The structure connection ∇_λ is defined as:

$$\nabla_{\lambda, X} Y = \nabla_{0, X} Y + \lambda X \circ Y.$$

By the fundamental theorem (cf. Theorem 10.7), $\mathcal{M}$ is Frobenius iff ∇_λ is flat for all λ. Since:

- ∇_0 is flat (given).
- The λ^2-curvature vanishes by commutativity/associativity.

we need only ensure vanishing of the λ-linear curvature terms. The curvature condition is:

$$[\nabla_{\lambda,\mathbf{e}_i}, \nabla_{\lambda,\mathbf{e}_j}](\mathbf{e}_k) = \nabla_{\lambda,[\mathbf{e}_i,\mathbf{e}_j]}(\mathbf{e}_k) \tag{10.10}$$

Expanding the left side:

$$(\nabla_{0,\mathbf{e}_i} + \lambda \mathbf{e}_i \circ)(\nabla_{0,\mathbf{e}_j} + \lambda \mathbf{e}_j \circ)(\mathbf{e}_k) - (i \leftrightarrow j)$$

Substitute $\nabla_{0,\mathbf{e}_j}\mathbf{e}_k = \sum_m \Gamma^m_{jk} e_m$ and $\mathbf{e}_j \circ \mathbf{e}_k = \delta_{jk}\mathbf{e}_j$:

$$= \nabla_{0,\mathbf{e}_i}\left(\sum_m \Gamma^m_{jk} e_m + \lambda\delta_{jk}\mathbf{e}_j\right) + \lambda \mathbf{e}_i \circ \left(\sum_m \Gamma^m_{jk} e_m + \lambda\delta_{jk}\mathbf{e}_j\right) - (i \leftrightarrow j)$$

Distribute and simplify:

$$= \sum_m \left[\mathbf{e}_i(\Gamma^m_{jk}) e_m + \Gamma^m_{jk} \nabla_{0,\mathbf{e}_i} e_m\right] + \lambda\delta_{jk}\nabla_{0,\mathbf{e}_i}\mathbf{e}_j + \lambda \sum_m \Gamma^m_{jk}\delta_{im}\mathbf{e}_i + \lambda^2 \delta_{jk}\delta_{ij}\mathbf{e}_i - (i \leftrightarrow j)$$

The λ-linear part is:

$$\sum_m \mathbf{e}_i(\Gamma^m_{jk})\mathbf{e}_m + \sum_m \Gamma^m_{jk}\Gamma^\ell_{im}\mathbf{e}_\ell + \lambda\delta_{jk}\Gamma^\ell_{ij}\mathbf{e}_\ell + \lambda\Gamma^i_{jk}\mathbf{e}_i - (i \leftrightarrow j) \tag{10.11}$$

The right side of (10.10) is:

$$\nabla_{\lambda,[\mathbf{e}_i,\mathbf{e}_j]}\mathbf{e}_k = \sum_m c^m_{ij}\left(\nabla_{0,e_m}\mathbf{e}_k + \lambda\delta_{mk}\mathbf{e}_m\right)$$

where $[\mathbf{e}_i, \mathbf{e}_j] = \sum_m c^m_{ij}\mathbf{e}_m$. Its λ-linear part is $\sum_m c^m_{ij}\delta_{mk}\mathbf{e}_m$.

Under condition (1) ($c^m_{ij} = 0$), equating λ-coefficients in (10.10) yields:

$$\delta_{jk}\Gamma^\ell_{ij} e_\ell + \Gamma^i_{jk}\mathbf{e}_i - \delta_{ik}\Gamma^\ell_{ji} e_\ell - \Gamma^j_{ik}\mathbf{e}_j = 0 \tag{10.12}$$

The Levi-Civita connection coefficients for the diagonal metric $g(\mathbf{e}_i, \mathbf{e}_j) = \eta_i\delta_{ij}$ are:

$$\Gamma^k_{ij} = \begin{cases} \frac{1}{2\eta_i}\mathbf{e}_j(\eta_i) & i = k \neq j \\ -\frac{1}{2\eta_j}\mathbf{e}_i(\eta_j) & j = k \neq i \\ \frac{1}{2\eta_i}\mathbf{e}_i(\eta_i) & i = j = k \\ 0 & \text{otherwise} \end{cases}$$

Under condition (2) ($\mathbf{e}_i(\eta_j) = \mathbf{e}_j(\eta_i)$), substitution into (10.12) shows all components vanish. The reverse implication follows from the existence of potential Φ in canonical coordinates.

The function η is the *metric potential*, related to the Frobenius potential Φ via:

$$\eta_i = \frac{\partial \eta}{\partial u^i}, \quad g_{ij} = \frac{\partial^2 \Phi}{\partial u^i \partial u^j}.$$

Canonical coordinates $\{u^i\}$ are unique up to permutation and additive constants. The closed co-identity $\mathcal{E} = d\eta$ reflects the Hessian nature of the metric.

Consider the Lie bracket $[\mathbf{e}_i, \mathbf{e}_j] = \sum_q f_{ij}^q \mathbf{e}_q$. The λ-linear term in $\nabla_{\lambda,[\mathbf{e}_i,\mathbf{e}_j]}(\mathbf{e}_k)$ is:

$$\lambda \sum_q f_{ij}^q (\mathbf{e}_q \circ \mathbf{e}_k) = \lambda \sum_q f_{ij}^q \delta_{qk} \mathbf{e}_k = \lambda f_{ij}^k \mathbf{e}_k.$$

Since the λ-linear terms in Eq. 10.11 vanish identically, and $\mathcal{M}$ carries a Frobenius structure (so ∇_λ is flat), we must have:

$$f_{ij}^k = 0 \quad \forall i, j, k.$$

Thus $[\mathbf{e}_i, \mathbf{e}_j] = 0$, verifying the existence of canonical coordinates $\{u^i\}$.

For the Levi-Civita connection of $g = \sum_i \eta_i (du^i)^2$, the Christoffel symbols are:

$$\Gamma_{ij}^k = \begin{cases} \frac{1}{2}\eta_i^{-1}\partial_i \eta_i & i = j = k \\ \frac{1}{2}\eta_i^{-1}\partial_j \eta_i & i = k \neq j \\ -\frac{1}{2}\eta_j^{-1}\partial_i \eta_j & j = k \neq i \\ 0 & \text{otherwise} \end{cases}.$$

The covariant derivatives are:

$$\nabla_{\mathbf{e}_i} \mathbf{e}_j = \begin{cases} \sum_{k \neq i} \left(-\frac{1}{2}\eta_k^{-1}\partial_k \eta_i\right) \mathbf{e}_k & i = j \\ \frac{1}{2}\eta_i^{-1}\partial_j \eta_i \mathbf{e}_i + \frac{1}{2}\eta_j^{-1}\partial_i \eta_j \mathbf{e}_j & i \neq j \end{cases}.$$

The vanishing curvature condition implies for all i, j, k:

$$\nabla_i(\mathbf{e}_j \circ \mathbf{e}_k) + \mathbf{e}_i \circ \nabla_j(\mathbf{e}_k) = \nabla_j(\mathbf{e}_i \circ \mathbf{e}_k) + \mathbf{e}_j \circ \nabla_i(\mathbf{e}_k) \tag{10.13}$$

Using $\mathbf{e}_j \circ \mathbf{e}_k = \delta_{jk}\mathbf{e}_j$, this simplifies to:

$$\delta_{jk}\nabla_i \mathbf{e}_j + \mathbf{e}_i \circ \nabla_j \mathbf{e}_k = \delta_{ik}\nabla_j \mathbf{e}_i + \mathbf{e}_j \circ \nabla_i \mathbf{e}_k.$$

- When $i = j = k$: holds identically.
- When $i = j \neq k$: reduces to $\nabla_i \mathbf{e}_k = 0$, which requires $\partial_i \eta_k = \partial_k \eta_i = 0$.
- When $i \neq j = k$: yields $\partial_i \eta_j = \partial_j \eta_i$.

By symmetry, $\partial_i \eta_j = \partial_j \eta_i$ for all pairs, implying locally there exists η such that $\eta_i = \partial_i \eta$. This completes the proof. □

10.7.7 Structure Connection and Curvature

We now turn to an analysis of the geometric properties of the pencil of connections ∇_λ, particularly in relation to the structures of a pre-Frobenius and Frobenius manifold.

Let ∇_λ denote the structure connection associated with the pre-Frobenius manifold $(\mathcal{M}, g, A)$. The curvature of this connection satisfies a quadratic relation in the parameter λ, taking the form:

$$\nabla_\lambda^2 = R_1\lambda^2 + R_2\lambda + R_3,$$

where R_1, R_2, R_3 are curvature terms determined by the pre-Frobenius structure. A key observation is that the term R_3 coincide with λ_0^2 which satisfies $\lambda_0^2 = 0$, leading to the simplification:

$$\nabla_\lambda^2 = R_2\lambda^2 + R_1\lambda.$$

This relation gives rise to the following fundamental theorem, characterizing the Frobenius condition in terms of the vanishing of specific curvature terms.

Theorem 10.11 *Let ∇_λ be the structure connection associated with the pre-Frobenius manifold $(\mathcal{M}, g, A)$. Then:*

- *The vanishing of R_1, i.e. $R_1 = 0$ if and only if $(\mathcal{M}, g, A)$ is equivalent to the potentiality of (M, g, A);*
- *The vanishing of R_2, i.e. $R_2 = 0$ is equivalent to the associativity of $(\mathcal{M}, g, A)$.*

Thus, the manifold (M, g, A) is Frobenius satisfies the full Frobenius condition if and only if the pencil of connections ∇_λ is flat.

Exercise 10.4 Write a proof.

10.7.8 Semisimple Frobenius Manifolds and Web Structures

10.7.8.1 Geometric Webs and Foliations

In this subsection, we discuss the construction of flat 3-webs via semisimple 3-dimensional Frobenius manifolds and provide a geometric interpretation of the Chern connection associated with these webs. We show that these webs are biholomorphic to the characteristic webs on the solutions of the corresponding associativity equations. Furthermore, these webs are hexagonal and possess at least one infinitesimal symmetry at each singular point.

Definition 10.9 (Fibration) A **fibration** $\lambda = (Y, X, \pi)$ consists of the following properties:

- a pair of differentiable manifolds Y (the total space, $\dim Y = m$) and X (the base space, $\dim X = n$) with $m > n$;
- a surjective submersion $\pi : Y \to X$ such that:

1. For all $x \in X, \pi^{-1}(x)$ is a submanifold diffeomorphic to the fiber F ($\dim F = m - n$).
2. For all $x \in X$, there exists a neighborhood $U_x \subset X$ with $\pi^{-1}(U_x) \cong U_x \times F$ (this is a local trivialization).

Given our interest in the *local* structure of such manifolds we can often think of them as connected domains of a Euclidean space of the same dimension.

The above definition allows us to introduce the notion of leaves.

Definition 10.10 (Foliation) Let Y be a smooth manifold of dimension m. A **k-dimensional foliation** on Y (where $\dim Y = m$ and $k \leq m$) is a decomposition of Y into a family of connected subsets $\{L_\alpha\}$ called *leaves*. Note that each point of Y lies on exactly one leaf. Eeach leaf L_α is required to be an *immersed* submanifold of dimension k.

The foliation is determined by a smooth k-dimensional distribution $\theta \subset \mathcal{T}Y$, assigning to each point $p \in Y$ a k-dimensional subspace $\theta_p \subset \mathcal{T}_pY$. This distribution is *integrable* in the sense of the Frobenius theorem: if X and Y are smooth vector fields tangent to θ, then their Lie bracket $[X, Y]$ is also tangent to θ.

10.7.8.2 Web Structures from Frobenius Manifolds

A **3-web** denoted by $W(3, 3, 1)$ consists of three ingredients:

- **3 foliations** (that is families of curves) on a 2-dimensional surface,
- each foliation has **3 distinct directions** at every point,
- **1 condition** (e.g., transversality, curvature constraints).

Let us give a local description of this. Let (x, y) be local coordinates on a surface. The foliations are given by:

$$\begin{cases} \mathcal{F}_1 : & \omega_1 = 0, \\ \mathcal{F}_2 : & \omega_2 = 0, \\ \mathcal{F}_3 : & \omega_3 = 0, \end{cases}$$

where $\omega_i = a_i(x, y)dx + b_i(x, y)dy$ are 1-forms. The **web condition** requires that $\omega_1 \wedge \omega_2 \wedge \omega_3 = 0$ (which correponds to linear dependence).

An important object occurring here is the **Chern connection** that we will write as ∇^W. This is the unique torsion-free affine connection preserving the web structure. It is constructed as follows.

1. **Choose an adapted coframe** (θ_1, θ_2) such that:

$$\omega_1 = \theta_1, \quad \omega_2 = \theta_2, \quad \omega_3 = \theta_1 + \theta_2.$$

2. **Impose the torsion-free condition**:

$$d\theta_i = \sum_j \Gamma^i_{jk}\theta_j \wedge \theta_k, \quad \Gamma^i_{jk} = \Gamma^i_{kj}.$$

3. Add the **web invariance** condition:

$$\nabla\omega_i \equiv 0 \pmod{\omega_i},$$

meaning the connection preserves each foliation.

The connection coefficients Γ^i_{jk} are determined by:

$$\begin{cases} d\theta_1 = \Gamma^1_{12}\theta_1 \wedge \theta_2, \\ d\theta_2 = \Gamma^2_{12}\theta_1 \wedge \theta_2. \end{cases}$$

In particular, the **curvature** of the Chern connection is given by:

$$\Omega = d\Gamma^1_{12} \wedge \theta_1 + d\Gamma^2_{12} \wedge \theta_2.$$

We discuss its geometric interpretation in terms of webs. Concerning **flatness**, if $\Omega = 0$, the web is **hexagonal** (which means the web is locally equivalent to three families of parallel lines). However, if we are in a non-flat case then the curvature measures the **non-associativity** of the web structure.

We now discuss an example of a planar 3-Web. For the standard 3-web in $\mathbb{R}^2$ given by:

$$\omega_1 = dx, \quad \omega_2 = dy, \quad \omega_3 = dx + dy,$$

the Chern connection is the **standard Euclidean connection** ($\Gamma^i_{jk} = 0$).

Let $(\mathcal{M}, g, \circ)$ be a 3D semisimple Frobenius manifold with canonical coordinates (u^1, u^2, u^3).

The idempotent fields $\mathbf{e}_i = \partial/\partial u^i$ define three pairwise transverse foliations:

$$\mathcal{F}_i : \text{ integral manifolds of } \ker(du^j) \cap \ker(du^k) \quad (i, j, k \text{ distinct})$$

These form a **3-web** $W(3, 3, 1)$ on $\mathcal{M}$. The Chern connection ∇^W of this web coincides with the Levi-Civita connection of g.

Corollary 10.1 (***Web Properties***) *The characteristic web of a 3D semisimple Frobenius manifold satisfies:*

1. ***Hexagonality***: *the Chern curvature vanishes (∇^W flat).*
2. ***Infinitesimal Symmetry***: *there exists and Euler field E with $\mathcal{L}_E(\mathcal{F}_i) = \mu_i \mathcal{F}_i$.*

10.7.8.3 Some Results on Characterization via Cubic ODEs

Consider the first-order differential equation

$$F(x, y, p) = p^3 + a(x, y)p^2 + b(x, y)p + c(x, y) = 0, \qquad p = \frac{\mathrm{d}y}{\mathrm{d}x}.$$

We regard (x, y, p) as coordinates in $\mathbb{R}^3$, and the equation $F(x, y, p) = 0$ defines a smooth surface $\mathcal{S} \subset \mathbb{R}^3$.

A solution $y = y(x)$ of the differential equation corresponds to a curve $x \mapsto (x, y(x))$ in the (x, y)-plane. This curve lifts naturally to the graph

$$x \mapsto \big(x, y(x), y'(x)\big)$$

in $\mathbb{R}^3$, which lies in $\mathcal{S}$. The lifted curve is tangent to the *contact distribution*

$$\xi = \ker(\alpha), \qquad \alpha = dy - p\,dx.$$

The *critical set* C is defined as the locus of points $m = (x_0, y_0, p_0) \in \mathcal{S}$ where both

$$F(x_0, y_0, p_0) = 0 \ \text{ and } \ \frac{\partial F}{\partial p}(x_0, y_0, p_0) = 0$$

are satisfied. Geometrically, C corresponds to singular points of the ODE on the surface $\mathcal{S}$. The *critical set* is the curve in $\mathcal{S}$ defined by

$$C = \{(x, y, p) \in \mathcal{S} \mid \tfrac{\partial F}{\partial p}(x, y, p) = 0\}.$$

At points of C, the differential equation admits multiple solutions passing through the same point in the (x, y)-plane; in other words, distinct branches of solutions coalesce.

This leads us to stating the following theorem.

Theorem 10.12 *Assume that the implicit cubic ODE*

$$p^3 + a(x, y)p^2 + b(x, y)p + c(x, y) = 0$$

has a flat web of solutions (i.e., the Chern connection of its 3-web is flat) and that it satisfies the regularity condition at $m = (x_0, y_0, z_0) \in C \subset \mathcal{S}$ (non-degenerate singularity). Then there exists a local diffeomorphism around $\pi(m) = (x_0, y_0)$ that reduces the ODE to:

- $p^3 + px - y = 0$, *if* p_0 *is a triple root and* $\mathcal{C}$ *is Legendrian.*[8]
- $p^3 + 2xp + y = 0$, *if* p_0 *is a triple root and* $\mathcal{C}$ *is not Legendrian.*
- $p^2 - y = 0$, *if* p_0 *is a double root and* $\mathcal{C}$ *is Legendrian.*
- $p^2 - x = 0$, *if* p_0 *is a double root and* $\mathcal{C}$ *is not Legendrian.*

Geometric interpretation The first normal form corresponds to quantum cohomology of $\mathbb{CP}^2$ (Frobenius potential $\Phi = \frac{1}{2}u_1^2u_3 + \frac{1}{2}u_1u_2^2 - \frac{1}{48}u_3^4$), while the second arises in A_3-singularity theory. Both define hexagonal webs via their characteristic varieties. □

10.8 Solutions of Exercises

Exercise 10.1.

Proof Define the *associator*:

$$\mathsf{A}(X, Y, Z) = (X \circ Y) \circ Z - X \circ (Y \circ Z) = \nabla_{\nabla_X Y} Z - \nabla_X(\nabla_Y Z).$$

The pre-Lie condition requires symmetry:

$$\mathsf{A}(X, Y, Z) - \mathsf{A}(Y, X, Z) = 0.$$

Expanding the left-hand side:

$$\mathsf{A}(X, Y, Z) - \mathsf{A}(Y, X, Z) = \left(\nabla_{\nabla_X Y} Z - \nabla_X \nabla_Y Z\right) - \left(\nabla_{\nabla_Y X} Z - \nabla_Y \nabla_X Z\right).$$

Rearranging terms:

$$= \nabla_{\nabla_X Y} Z - \nabla_{\nabla_Y X} Z - (\nabla_X \nabla_Y Z - \nabla_Y \nabla_X Z).$$

By the torsion-free property

$$\nabla_X Y - \nabla_Y X = [X, Y],$$

[8] Let (M^{2n+1}, ξ) be a **contact manifold**, where $\xi = \ker \alpha$ for a 1-form α satisfying

$$\alpha \wedge (d\alpha)^n \neq 0.$$

A **Legendrean root** is a geometric object (e.g., a distribution, submanifold, or integral curve) satisfying:

A submanifold $L \subset M$ is *Legendrian* if $\dim L = n$ and $TL \subset \xi$ (i.e., $\alpha|_L = 0$ and $d\alpha|_L = 0$). The "root" is a minimal/maximal Legendrian subspace governing deformations or singularities.

the first two terms simplify to:

$$\nabla_{\nabla_X Y} Z - \nabla_{\nabla_Y X} Z = \nabla_{[X,Y]} Z.$$

The remaining terms are:

$$-(\nabla_X \nabla_Y Z - \nabla_Y \nabla_X Z) = -[\nabla_X, \nabla_Y] Z.$$

By flatness:

$$[\nabla_X, \nabla_Y] Z = \nabla_X \nabla_Y Z - \nabla_Y \nabla_X Z = \nabla_{[X,Y]} Z.$$

Thus:

$$\mathsf{A}(X, Y, Z) - \mathsf{A}(Y, X, Z) = \nabla_{[X,Y]} Z - \nabla_{[X,Y]} Z = 0.$$

This confirms the pre-Lie algebra structure. □

Exercise 10.3.

It is easily shown by using the paper of Calabi (1954) [Ca54] and the definition of a potential pre-Frobenius manifold.

Exercise 10.4.

The proof proceeds in two stages. **Part 1.** The coefficient R_1 of λ in the commutator

$$\left[\nabla_{0,\partial_a} + \lambda \partial_a \circ,\ \nabla_{0,\partial_b} + \lambda \partial_b \circ\right](\partial_c)$$

vanishes if and only if

$$\partial_a A^e_{bc} = (-1)^{ab} \partial_b A^e_{ac} \quad \forall a, b, c, e. \tag{10.14}$$

In covariant form (assuming constant metric), this becomes

$$\partial_a A_{bcd} = (-1)^{ab} \partial_b A_{acd} \tag{10.15}$$

where A_{bcd} is the fully covariant tensor.

Note: The factor $(-1)^{ab}$ arises naturally in $\mathbb{Z}_2$-graded settings but is retained here for generality. It encodes the sign rule for graded commutativity when indices a, b represent parity degrees.

The symmetry $A(X, Y, Z) = (XYZ)\Phi$ implies $A_{acd} = (-1)^{a(b+c)} A_{bcd}$, but we proceed without this assumption. From (10.15), for fixed c, d, consider the 1-form

$$\omega_{cd} = \sum_b A_{bcd} dx^b.$$

Its exterior derivative is

$$d\omega_{cd} = \sum_{a<b} \left(\partial_a A_{bcd} - (-1)^{\tilde{a}\tilde{b}} \partial_b A_{acd}\right) dx^a \wedge dx^b.$$

Substituting (10.15) shows $d\omega_{cd} = 0$. By the Poincaré lemma, ω_{cd} is locally exact: there exist functions B_{cd} such that

$$A_{bcd} = \partial_b B_{cd}. \tag{10.16}$$

Symmetry constraint: Graded symmetry $A_{bcd} = (-1)^{bc} A_{cbd}$ implies

$$\partial_b B_{cd} = (-1)^{bc} \partial_c B_{bd}.$$

For fixed d, consider the 1-form $\gamma_d = \sum_c B_{cd} dx^c$. Its exterior derivative vanishes:

$$d\gamma_d = \sum_{a<c} \left(\partial_a B_{cd} - (-1)^{\tilde{a}\tilde{c}} \partial_c B_{ad} \right) dx^a \wedge dx^c = 0$$

by the symmetry condition. Thus γ_d is locally exact: there exist functions C_d such that

$$B_{cd} = \partial_c C_d.$$

Substituting into (10.16) gives $A_{bcd} = \partial_b \partial_c C_d$. Graded symmetry now implies

$$\partial_b \partial_c C_d = (-1)^{b(c+d)} \partial_c \partial_d C_b,$$

which by equality of mixed partials $\partial_i \partial_j = (-1)^{\tilde{i}\tilde{j}} \partial_j \partial_i$ yields $C_d = \partial_d \Phi$ for some Φ. Thus

$$A_{bcd} = \partial_b \partial_c \partial_d \Phi.$$

Part 2. The λ^2 coefficient in $[\nabla_{\lambda,X}, \nabla_{\lambda,Y}](Z)$ is

$$R_{2,XY}(Z) = X \circ (Y \circ Z) - (-1)^{\tilde{X}\tilde{Y}} Y \circ (X \circ Z).$$

If $\circ$ is associative, then $R_2 = 0$ by commutativity. Conversely, if $R_2 = 0$, then

$$X \circ (Y \circ Z) = (-1)^{\tilde{X}\tilde{Y}} Y \circ (X \circ Z) = (-1)^{\tilde{X}(\tilde{Y}+\tilde{Z})} (Y \circ X) \circ Z,$$

implying associativity. □

Unveiling the Hidden Geometry of Statistical Manifolds

11

In this chapter, and the following one, we reveal the intricate (and often hidden!) geometric structures underlying statistical manifolds. In particular, we revisit certain geometric concepts that have long been overlooked, such as m-pairs, and demonstrate their relevance in this context. By bringing these ideas back into focus, we aim to provide a deeper understanding of the rich interplay between geometry and statistical structures.

11.1 Classical Projective Geometry

Projective geometry is a non-Euclidean geometry that extends the concepts of Euclidean geometry by introducing **points at infinity**, where parallel lines intersect. This provides a more unified geometric framework.

In particular, a first striking feature is that there are **no parallel lines:** any two distinct lines in the projective plane $\mathbb{P}^2$ intersect at exactly one point (possibly at infinity).

A second important property is the existence of a certain **duality principle.** Every statement about points and lines has a dual statement where points and lines are interchanged.

Projective points are defined via **homogeneous coordinates.** In particular a point in a projective space is represented by an equivalence class $[x_0 : \cdots : x_n]$ where scaling by a scalar $\lambda \neq 0$ does not change that projective point. We will go back to this notion more precisely below.

Finally, projective spaces are naturally endowed with **projective transformations.** These are invertible linear maps on homogeneous coordinates that preserve incidence relations.

© The Author(s), under exclusive license to Springer Nature Switzerland AG 2026

N. C. Combe et al., *Exploring Information Geometry*, Frontiers in Mathematics,
https://doi.org/10.1007/978-3-032-23054-6_11

11.1.1 Projective Space

The n-**dimensional projective space** $\mathbb{P}^n$ over a field $\mathbb{K}$ is defined as:

$$\mathbb{KP}^n = \left(\mathbb{K}^{n+1} \setminus \{0\}\right) / \sim$$

where the equivalence relation is given by $(x_0, \ldots, x_n) \sim (y_0, \ldots, y_n)$ if there exists a non vanishing scalar $\lambda \neq 0$ such that $y_i = \lambda x_i$ for all i. In a projective space $\mathbb{KP}^n$, the homogeneous coordinates of a point will be denoted as $[X_0 : X_1 : \cdots : X_n]$.

For instance, in a projective plane $\mathbb{KP}^2$, the homogeneous coordinates of a point will be denoted as $[X_0 : X_1 : X_2]$.

Lines in $\mathbb{KP}^2$ are defined by homogeneous equations of type $aX + bY + cZ = 0$ where coefficients lie in $a, b, c \in \mathbb{K}$. The **line at infinity** in $\mathbb{KP}^2$ is given by $Z = 0$.

Two distinct lines in $\mathbb{P}^2$ given by the equations

$$\begin{cases} aX + bY + cZ = 0 \\ dX + eY + fZ = 0 \end{cases}$$

always intersect at a point (found via cross product in homogeneous coordinates. The cross product of the coefficient vectors of the lines provides homogeneous coordinates for this intersection point.). More precisely, if we consider two distinct lines in the projective plane $\mathbb{P}^2$, given in homogeneous coordinates by

$$aX + bY + cZ = 0 \quad \text{and} \quad dX + eY + fZ = 0.$$

Their coefficient vectors

$$U = (a, b, c) \quad \text{and} \quad V = (d, e, f)$$

are linearly independent, reflecting the fact that the lines are distinct.

A classical construction in projective geometry asserts that the intersection point of these two lines may be obtained via the cross product:

$$U \times V = (bf - ce,\ cd - af,\ ae - bd).$$

Since U and V are linearly independent, this vector is nonzero. Moreover, a straightforward computation verifies that $(X, Y, Z) = U \times V$ satisfies both line equations:

$$a(bf - ce) + b(cd - af) + c(ae - bd) = 0,$$
$$d(bf - ce) + e(cd - af) + f(ae - bd) = 0.$$

Thus, the projective point

$$[X : Y : Z] = [bf - ce : cd - af : ae - bd]$$

lies on both lines.

From the linear independence of U and V, it follows that the 2×3 matrix

$$\begin{pmatrix} a & b & c \\ d & e & f \end{pmatrix}$$

has rank 2, so the solution space of the system is one-dimensional. Consequently, every solution is a scalar multiple of $U \times V$, and the above coordinates represent the unique intersection point in $\mathbb{P}^2$.

If, on the other hand, the lines coincide, U and V are linearly dependent and the cross product vanishes; in that degenerate case, there is no unique intersection point. In what follows, we work under the standing assumption that the lines are distinct, so that the cross product furnishes a well-defined point of intersection.

Remark 11.1 The **projective complex space** $\mathbb{CP}^n$ and the **projective real space** $\mathbb{RP}^n$ are fundamental constructions in geometry and topology, but they differ significantly due to their underlying fields ($\mathbb{C}$ vs. $\mathbb{R}$). Below is a detailed comparison.

- $\mathbb{CP}^n$ is the space of all complex lines through the origin in $\mathbb{C}^{n+1}$:

$$\mathbb{CP}^n = \left(\mathbb{C}^{n+1} \setminus \{0\}\right) / \sim, \quad \text{where } \mathbf{u} \sim \mathbf{v} \text{ if } \mathbf{u} = \lambda \mathbf{v} \text{ for some } \lambda \in \mathbb{C}^*.$$

- $\mathbb{RP}^n$ is the space of all real lines through the origin in $\mathbb{R}^{n+1}$:

$$\mathbb{RP}^n = \left(\mathbb{R}^{n+1} \setminus \{0\}\right) / \sim, \quad \text{where } \mathbf{u} \sim \mathbf{v} \text{ if } \mathbf{u} = \lambda \mathbf{v} \text{ for some } \lambda \in \mathbb{R}^*.$$

The topology and geometry for $\mathbb{CP}^n$ and $\mathbb{RP}^n$ turns out to be deeply modified, as we highlight below (Table 11.1).

- $\mathbb{CP}^n$ is a **compact complex manifold** with a natural Kähler metric, making it a fundamental object in algebraic and complex geometry.
- $\mathbb{RP}^n$ is a **compact real manifold**, but its topology is more constrained (e.g., non-orientable for even n).
- $\mathbb{CP}^n$ is **simply connected**, whereas $\mathbb{RP}^n$ has fundamental group $\mathbb{Z}_2$ for $n \geq 2$.
- The cohomology of $\mathbb{CP}^n$ is $\mathbb{Z}$ in even degrees, while $\mathbb{RP}^n$ has $\mathbb{Z}_2$ coefficients in odd dimensions (for odd n).

Table 11.1 Comparison of $\mathbb{CP}^n$ and $\mathbb{RP}^n$

Property	$\mathbb{CP}^n$	$\mathbb{RP}^n$
Underlying field	$\mathbb{C}$	$\mathbb{R}$
Dimension	Complex n (Real $2n$)	Real n
Manifold type	Complex, Kähler	Real, Riemannian
Orientability	Always orientable	Orientable only if n is odd
Simply connected?	Yes	No ($\pi_1 = \mathbb{Z}_2$ for $n \geq 2$)
Homology	$\mathbb{Z}$ in even dims up to $2n$	$\mathbb{Z}_2$ in odd dims (for odd n)
Example ($n = 1$)	Riemann sphere ($\cong S^2$)	Circle ($\cong S^1$)

Exercise 11.1 Part 1: Distinct Lines. Find the intersection point of the following lines:

$$L_1: \quad X + 2Y - 3Z = 0, \qquad L_2: \quad 2X - Y + Z = 0.$$

Use the cross product method and express the intersection point in homogeneous coordinates $[X : Y : Z]$.

Part 2: Coincident Lines. Show that the lines

$$L_1: \quad 3X - Y + 4Z = 0, \qquad L_2: \quad 6X - 2Y + 8Z = 0$$

are coincident, and explain why the cross product method fails to produce a well-defined intersection point in this case.

Hint: Recall that the cross product of two linearly dependent vectors is zero.

11.1.1.1 Intersection and Multiplicty

In projective geometry, it is common to be considering a pair of geometric objects in the projective space and to be checking whether they intersect or not. In particular, if the intersection is non-empty then intersection multiplicities appear. The definitions of intersection multiplicity and order of contact are algebraic and apply to any of the fields $\mathbb{R}$ or $\mathbb{C}$.

Assume that two algebraic curves $C_1, C_2 \subset \mathbb{KP}^2$ intersect at a point P. Then, the multiplicity of their intersection can be one of the following:

- A transverse intersection, $m = 1$ (simple crossing).
- A tangent intersection: $m = 2$ (first order contact).
- A higher order contact: $m \geq 3$.

However, its geometric interpretation and applicability depend on the field's properties and this is a delicate question to consider. We develop more concretely an example in the complex case.

Consider a point $O \in \mathbb{CP}^2$ and an algebraic curve C passing through O. Let $\{L_t\}_{t\in\mathbb{P}^1}$ be a pencil of lines through O. For each line L_t, we denote by

$$\operatorname{ord}\big((L_t \cap C)_O\big)$$

the intersection multiplicity of L_t and C at O, or simply write m_t for short.

We have the basic inequality

$$m_t \;\geq\; \mu_O(C),$$

where $\mu_O(C)$ denotes the multiplicity of C at O. Equality

$$m_t \;=\; \mu_O(C)$$

holds for generic $t \in \mathbb{P}^1$. A strict inequality

$$m_t \;>\; \mu_O(C)$$

occurs if and only if L_t is tangent to C at O. In this sense, m_t measures the *order of contact* between the line L_t and the curve C at the point O, reflecting the degree of tangency.

Example: Conic-Line intersection We consider an elementary example given by a conic-line intersection in $\mathbb{CP}^2$. Consider the conic $\{C : XZ - Y^2 = 0\}$ and the line $\{L : Y = 0\}$ in $\mathbb{CP}^n$. Then, the intersection of those two objects is given by

$$\begin{aligned} C \cap L &= \{[X : Y : Z] \mid Y = 0, XZ = 0\} \\ &= \{[1 : 0 : 0],\ [0 : 0 : 1]\}. \end{aligned}$$

Now, let us consider L' to be the tangent line $X = 0$. Then, the intersection of C with L' is given by:

$$\begin{aligned} C \cap L' &= \{[0 : Y : Z] \mid -Y^2 = 0\} \\ &= \{[0 : 0 : 1]\} \text{ (with multiplicity 2)} \end{aligned}$$

Therefore, the point $[0 : 0 : 1]$ has multiplicity 2.

Example: Cubic-Line intersection Let $O = [0 : 0 : 1]$ and consider the curve $\{C : Y^2Z - X^3 = 0, \}$ which has a cusp at O; in particular, $\mu_O(C) = 2$[1]. The pencil of lines through O is given by

[1] We examine the curve at the origin. The *multiplicity* of a curve at a point is the smallest integer m such that the Taylor expansion of its defining polynomial contains a non-zero term of total degree m. Here the presence of a non-zero second derivative indicates that the lowest-degree non-zero term in the Taylor expansion is quadratic. Since $\mu_O(C) > 1$, the point O is singular.

$$L_t : Y = tX, \quad t \in \mathbb{P}^1,$$

working in affine coordinates with $Z = 1$. Substituting $Y = tX$ into the equation of C yields

$$(tX)^2 \cdot 1 - X^3 \;=\; X^2(t^2 - X) \;=\; 0.$$

Thus, the intersections are given by $X = 0$ (with multiplicity 2) and $X = t^2$ (with multiplicity 1). At O $(X = 0)$, the intersection multiplicity is $m_t = 2$ for all t, since the equation has a double root at $X = 0$. This reflects the fact that every line L_t passes through the cusp, but none are tangent in the sense of increasing the multiplicity beyond $\mu_O(C) = 2$. Here,

$$m_t = \mu_O(C) \quad \text{for all } t,$$

so no line in the pencil is tangent to C at O.

Example: Order of Contact

To achieve $m_t > 2$, one must consider a curve with different local geometry. For instance, take the smooth parabola

$$C : Y - X^2 = 0$$

at O (so $\mu_O(C) = 1$), with the same pencil $L_t : Y = tX$. Substitution yields

$$tX - X^2 = X(t - X) = 0.$$

At O, the multiplicity is $m_t = 1$ for $t \neq 0$, corresponding to transverse intersection. However, for the special line $L_0 : Y = 0$, substitution gives $-X^2 = 0$, so

$$m_0 = 2 > 1.$$

Thus, L_0 is tangent to C at O, with order of contact 2.

In summary, for a given pencil $\{L_t\}_{t\in\mathbb{P}^1}$ through a point O on a curve C, the generic member has $m_t = \mu_O(C)$, while special members with $m_t > \mu_O(C)$ are precisely those tangent to C at O.

This is tightly related to the famous Bézout Theorem, that we state below for projective curves over complex numbers.

Theorem 11.1 (Bézout's Theorem) *Let $C_1, C_2 \subset \mathbb{CP}^2$ be curves of degrees d_1, d_2 with no common components. If they intersect at points $\{P_i\}_{i=1}^k$ with multiplicities m_i, then:*

$$\sum_{i=1}^{k} m_i = d_1 \cdot d_2$$

Proof The proof follows from computing the resultant of the defining polynomials and accounting for multiplicities in the vanishing locus. □

> When working over the field of real numbers, one must exercise care in applying Bézout's theorem. Although over the algebraic closure $\mathbb{C}$ the number of intersection points of two projective plane curves of degrees d and e (counted with multiplicity) is always $d \cdot e$, this equality may fail dramatically over $\mathbb{R}$.

The discrepancy stems from several intertwined phenomena.

(1) Insufficient Real Solutions. Over $\mathbb{R}$, two curves may have fewer *real* intersection points than the complex bound $d \cdot e$. A simple illustration is provided by two distinct conics ($d = e = 2$): although $d \cdot e = 4$, the actual number of real intersection points may be $0, 1, 2, 3,$ or 4, depending on the position of the curves. Likewise, a line ($d = 1$) and a circle ($e = 2$) can intersect in at most two real points, but may also be disjoint if the line does not meet the circle.

(2) Complex but Nonreal Solutions. Bézout's count is performed over $\mathbb{C}$; thus it includes solutions that are not necessarily real. The set of real intersection points is simply a subset of the complex intersection set, and many complex solutions appear in conjugate pairs without a real counterpart.

(3) Weaker Statements Over $\mathbb{R}$. The structural result—that if two curves meet in more than $d \cdot e$ points they must share a common component—remains valid over *any* field. However, over $\mathbb{R}$ the bound $d \cdot e$ represents merely a *maximum* number of real intersection points, and it need not be attained for a given configuration. Thus, while $d \cdot e$ remains an upper bound in the real setting, it is generically a loose one.

11.1.2 Normalization of Hypersurfaces and m-Pairs

Let us consider an object X_d, a d-dimensional hypersurface residing in an n-dimensional projective space $\mathbb{P}^n$, where the constraint $d \leq n$ holds.

Definition 11.1 We say that the hyersurface X_d is *normalized* if, for every point $p \in X_d$, we associate two distinct hyperplanes:

1. Normal of first type, P_I, is of dimension $(n - d)$ and intersects the tangent d-plane $T_p X_d$ at a unique point p.
2. Normal of the second type, P_{II}, is of dimension $(d - 1)$, situated within the d-plane $T_p X_d$, and does not pass through the point p.

Example 11.1 Consider the case where $d = 2$ and $n = 3$. This means we have a 2-dimensional surface X_2 within the 3-dimensional projective space $\mathbb{P}^3$.

At a point p on X_2:

- The tangent plane T_pX_2 is a 2-dimensional plane in the ambient $\mathbb{P}^3$ space.
- The normal of first type P_I would be a line that intersects the tangent plane T_pX_2 at the point p.
- The normal of second type P_{II} is another line that lies completely within the tangent plane T_pX_2 but does not contain the point p.

This embodies a duality, intrinsic to projective geometry. Notably, in the case where $d = n$, the hyperplane P_I identifies to the point p, and P_{II} becomes the $(n-1)$-dimensional surface devoid of the point p. This situation reflects the classical notion of duality in projective spaces, leading to the identification of X_n with the projective space $\mathbb{P}^n$.

Definition 11.2 We define an *m-pair* as a pair constituted of an m-plane and an $(n-m-1)$-plane in projective space.

More precisely, an m**-pair** is a pair consisting of:

1. An m**-plane**, i.e. a linear subspace of dimension m in an n-dimensional space.
2. An $(n-m-1)$**-plane**, i.e. a linear subspace of dimension $n-m-1$ in the same n-dimensional space.

These two planes are typically considered in the context of projective geometry or linear algebra, where they may satisfy certain geometric or algebraic relationships.

Moreover, normalized surfaces associated with an m-pair space have the following properties:

Lemma 11.1 *1. The collection of m-pairs forms a projective differentiable manifold.*
2. For any integer $m \geq 0$, a manifold of $m-$pairs contains two flat, affine, and symmetric connections.

11.1.3 Some Examples and Exercices

Let us take $n = 3$. We discuss 0-pairs and 1-pairs.

- A **0-pair** consists of a point (0-plane) and of a 2-plane. This can be identified with the projective space $\mathbb{P}^3$.

- A **1-pair** consists of a line (1-plane) and another line (1-plane). Two lines in 3 dimensional space can either intersect at a point, be parallel (not intersecting but lying on a common plane), or be skew (not intersecting and not parallel).

  Let us take $n = 4$.

- A **1-pair** consists of a line (1-plane) and of a 2-plane. In 4 dimensional space, a line and a plane can intersect at a point, not intersect at all, or intersect along a line if the line lies entirely within the plane.
- Concerning **2-pairs**, these are constituted from 2-planes. In 4 dimensional space, two planes can intersect at a point, along a line, or not intersect at all.

Exercise 11.2 Try to draw all m-pairs in the low dimensional cases.

In projective geometry, an m-pair can be used to describe configurations of points, lines, and planes at infinity. The concept of m-pairs is closely related to Grassmannians, which are spaces that parameterize all m-dimensional subspaces of an n-dimensional space. The study of m-pairs can be seen as a way to explore the relationships between different Grassmannians.

Exercise 11.3 Determine what kind of relation there exists between the notion of Grassmanianns and m-pairs.

Applications of m-pairs in Computer Vision

In computer vision, m-pairs can be used to model the relationship between different views of a scene. For example, in structure from motion, the relationship between 2D image planes (2-planes) and 3D space (3-planes) can be analyzed using the concept of m-pairs.

Exercise 11.4 Generate an example using your favorite software m-pairs modeling the relationship between different views of a scene.

The examples above illustrate how m-pairs can be applied in various contexts, from projective geometry to computer vision.

11.2 Generalization of Projective Spaces

We now turn our attention to *projective spaces over algebras*. Among these, the *paracomplex numbers* occupy a particularly intriguing position, not least because of their unexpected appearance in the field of information geometry. Whereas our earlier discussion remained closer to the realm of geometry, the next section moves decisively into algebra. In this algebraic landscape, paracomplex structures arise naturally in the study of certain statistical manifolds, offering an elegant framework for their analysis. An especially striking feature is the emergence of *mâŁ"pairs* in this context, linking the algebraic theory of paracomplex numbers to the geometric properties of information spaces.

11.2.1 Paracomplex Numbers

Definition 11.3 A **paracomplex number** is an element of the form:

$$z = x + \epsilon y,$$

where $x, y \in \mathbb{R}$ are real numbers, and ϵ is the paracomplex unit satisfying $\epsilon^2 = 1$.

The algebra of paracomplex numbers $\mathfrak{P}$ is a two-dimensional commutative algebra over the real numbers $\mathbb{R}$. It is isomorphic to the direct sum $\mathbb{R} \oplus \mathbb{R}$, and its multiplication rule is given by:

$$(x_1 + \epsilon y_1)(x_2 + \epsilon y_2) = (x_1x_2 + y_1y_2) + \epsilon(x_1y_2 + x_2y_1).$$

This algebra enjoys certain properties depicted below.

- **Idempotent Basis**: Paracomplex numbers can be expressed in terms of idempotent elements. Define:
$$e_+ = \frac{1+\epsilon}{2}, \quad e_- = \frac{1-\epsilon}{2}.$$
These elements satisfy $e_+^2 = e_+$, $e_-^2 = e_-$, and $e_+e_- = 0$. Any paracomplex number $z = x + \epsilon y$ can be written as:
$$z = (x+y)e_+ + (x-y)e_-.$$
- **Conjugation**: The paracomplex conjugate of $z = x + \epsilon y$ is defined as:
$$\bar{z} = x - \epsilon y.$$
- **Norm**: The norm of a paracomplex number $z = x + \epsilon y$ is given by:
$$\|z\| = z\bar{z} = x^2 - y^2.$$
Note that this norm is not positive definite, as it can take negative values.

Exercise 11.5 Show that $(a + a\epsilon) \cdot (b - b\epsilon) = 0$.

11.2.2 Module over a Paracomplex Algebra

A module over a paracomplex algebra generalizes the concept of a vector space, where the scalars are paracomplex numbers instead of real or complex numbers.

Definition 11.4 Let $\mathfrak{P}$ be the algebra of paracomplex numbers. A **module over** $\mathfrak{P}$ is an abelian group M together with a scalar multiplication:

$$\cdot : \mathfrak{P} \times M \to M,$$

satisfying the following properties for all $z, z_1, z_2 \in \mathfrak{P}$ and $m, m_1, m_2 \in M$:

1. $z \cdot (m_1 + m_2) = z \cdot m_1 + z \cdot m_2,$
2. $(z_1 + z_2) \cdot m = z_1 \cdot m + z_2 \cdot m\ ,$
3. $(z_1 z_2) \cdot m = z_1 \cdot (z_2 \cdot m)\ ,$
4. $1 \cdot m = m\ .$

Remark 11.2 The simplest example of a module over $\mathfrak{P}$ is $\mathfrak{P}^n$: the set of n-tuples of paracomplex numbers. Scalar multiplication is defined component-wise:

$$z \cdot (z_1, z_2, \ldots, z_n) = (zz_1, zz_2, \ldots, zz_n).$$

We outline some properties.

- **Decomposition into Real Submodules:** One of the remarkable features of modules over the paracomplex algebra $\mathfrak{P}$ is that they can be decomposed using the idempotent elements e_+ and e_-. Specifically, any module M over $\mathfrak{P}$ naturally splits into two real submodules, each corresponding to one of the idempotents. This decomposition provides a clear and intuitive way to study the structure of M by reducing questions about paracomplex modules to questions about ordinary real vector spaces. Concretely, we have

$$M = e_+ M \oplus e_- M,$$

  where $e_+ M$ and $e_- M$ are real vector spaces associated with the idempotents e_+ and e_-, respectively.
- Modules over paracomplex algebras arise naturally in a variety of mathematical and physical contexts. In geometry, they provide the algebraic underpinning for the study of *para-Hermitian* and *para-Kähler* manifolds, allowing one to capture the subtleties of split-signature structures in a concise algebraic language. In physics, paracomplex numbers and their associated modules play a significant role in the analysis of *supersymmetric* theories and *integrable systems*, offering a unifying framework in which algebraic and geometric properties can be studied simultaneously.

Table 11.2 Projective geometries over $\mathbb{R}$, $\mathbb{C}$, and $\mathfrak{P}$

Property	$\mathbb{RP}^n$	$\mathbb{CP}^n$	$\mathfrak{P}\mathbb{P}^n$
Underlying algebra	$\mathbb{R}$	$\mathbb{C}$	$\mathfrak{P}$
Real dimension	n	$2n$	$2n$
Manifold structure	Smooth	Complex	Singular
Metric signature	Definite	Positive-definite	Neutral (n, n)

11.2.3 Projective Paracomplex Space

Classical projective geometry comes in two celebrated flavors: real projective space $\mathbb{RP}^n$ and complex projective space $\mathbb{CP}^n$, obtained by taking lines through the origin in $\mathbb{R}^{n+1}$ and $\mathbb{C}^{n+1}$, respectively. A natural third sibling arises when one replaces the underlying field by the algebra of *paracomplex numbers* (also called *split-complex numbers*). The resulting space, the *projective paracomplex space*, retains the elegance of projective geometry while introducing a host of new and subtle phenomena.

In analogy with the classical constructions, we define

$$\mathfrak{P}\mathbb{P}^n = \left(\mathfrak{P}^{n+1} \setminus \{\text{zero divisors}\}\right) / \sim,$$

where $\mathbf{u} \sim \mathbf{v}$ if there exists a non-zero divisor $\lambda \in \mathfrak{P}$ such that $\mathbf{u} = \lambda \mathbf{v}$. The exclusion of zero divisors from the scalar group is a necessity: without it, the quotient would collapse in degenerate ways.

The local and global geometry of $\mathfrak{P}\mathbb{P}^n$ departs sharply from the real and complex cases:

- *Real Dimension*: since $\mathfrak{P} \cong \mathbb{R}^2$ as a vector space, $\mathfrak{P}\mathbb{P}^n$ has real dimension $2n$.
- *Topology*: the presence of zero divisors introduces singular strata, preventing $\mathfrak{P}\mathbb{P}^n$ from being a smooth manifold in the classical sense.
- *Idempotent Decomposition*: writing $e_\pm = \frac{1 \pm \epsilon}{2}$ yields $\mathfrak{P} = e_+\mathbb{R} \oplus e_-\mathbb{R}$, from which one sees that $\mathfrak{P}\mathbb{P}^n$ locally resembles two copies of $\mathbb{RP}^n$ glued along a singular set.
- *Geometry*: $\mathfrak{P}\mathbb{P}^n$ naturally carries a para-Hermitian structure, the split-signature analogue of the Kähler structure on $\mathbb{CP}^n$.

We compare this projective space to the ones encountered formerly (Table 11.2).

The geometry of $\mathfrak{P}\mathbb{P}^n$ has appeared–sometimes implicitly–in several settings:

- In *theoretical physics*, as a natural stage for models with neutral signature metrics, including certain Wick-rotated spacetimes.
- In the theory of *integrable systems* and *symmetric spaces*, where para-complex geometry provides a unifying language for structures of split type.

The projective paracomplex space offers a meeting ground between the rigidity of algebraic geometry and the flexibility of differential geometry in split signature. Its singularities, idempotent decomposition, and neutral metric structure distinguish it sharply from $\mathbb{RP}^n$ and $\mathbb{CP}^n$, and invite further exploration–both for their intrinsic beauty and for their role in mathematics and physics.

11.3 Solutions to Exercise

Exercise 11.1. **Part 1: Distinct Lines.** We are given

$$L_1: \quad X + 2Y - 3Z = 0, \qquad L_2: \quad 2X - Y + Z = 0.$$

The coefficient vectors are

$$(a, b, c) = (1, 2, -3), \qquad (d, e, f) = (2, -1, 1).$$

The cross product is

$$\begin{vmatrix} \mathbf{i} & \mathbf{j} & \mathbf{k} \\ 1 & 2 & -3 \\ 2 & -1 & 1 \end{vmatrix} = \mathbf{i}(2 \cdot 1 - (-3)(-1)) - \mathbf{j}(1 \cdot 1 - (-3) \cdot 2) + \mathbf{k}(1 \cdot (-1) - 2 \cdot 2).$$

Simplifying,

$$= \mathbf{i}(2 - 3) - \mathbf{j}(1 + 6) + \mathbf{k}(-1 - 4) = (-1, -7, -5).$$

Since homogeneous coordinates are defined up to nonzero scalar multiples we rather take the point $[1 : 7 : 5]$ which lies indeed on both lines as we can see:

$$L_1: \quad 1(1) + 2(7) - 3(5) = 1 + 14 - 15 = 0,$$
$$L_2: \quad 2(1) - 1(7) + 1(5) = 2 - 7 + 5 = 0.$$

Hence the lines intersect uniquely at $[1 : 7 : 5]$.

Part 2: Coincident Lines. We have

$$L_1: \quad 3X - Y + 4Z = 0, \qquad L_2: \quad 6X - 2Y + 8Z = 0,$$

with coefficient vectors

$$(a, b, c) = (3, -1, 4), \qquad (d, e, f) = (6, -2, 8).$$

The cross product is

$$\begin{vmatrix} \mathbf{i} & \mathbf{j} & \mathbf{k} \\ 3 & -1 & 4 \\ 6 & -2 & 8 \end{vmatrix} = \mathbf{i}((-1) \cdot 8 - 4 \cdot (-2)) - \mathbf{j}(3 \cdot 8 - 4 \cdot 6) + \mathbf{k}(3 \cdot (-2) - (-1) \cdot 6).$$

Simplifying,

$$= \mathbf{i}(-8 + 8) - \mathbf{j}(24 - 24) + \mathbf{k}(-6 + 6) = (0, 0, 0).$$

The zero vector does not define a point in $\mathbb{P}^2$.

Explanation: Since

$$6X - 2Y + 8Z = 2(3X - Y + 4Z),$$

the lines L_1 and L_2 are identical. Every point on L_1 lies on L_2, so there is no *unique* intersection point. The cross product method fails because the coefficient vectors are linearly dependent, resulting in the zero vector.

12 How Frobenius structures, Paracomplex Spaces Meets Caph

> Jordan algebras and Jordan logic —the logical framework derived from Jordan algebras—offer a tool for understanding the structure of quantum observables and their geometric properties, with significant implications for quantum information theory. Building on the foundational work of Chentsov–Morozova (1991) [MoCh91-1], these algebraic structures provide insights into the unique features of quantum mechanics, such as non-commutativity, measurement statistics, and information geometry.

The connection between the manifold of probability distributions (specifically exponential families), Vinberg cones, and Jordan algebras represents a profound synthesis of algebraic geometry, information theory, and convex analysis. This relationship extends classical results by Vinberg and others for instance [V63, FK94] as well as [CM20, Le94] revealing deep structures in statistical manifolds.

The theory of homogeneous convex cones admits a deep and elegant classification due to Vinberg, formulated through T*-algebras*—a generalization of formally real Jordan algebras. Recall that an open convex cone $\mathcal{C} \subset \mathbb{R}^n$ is said to be *homogeneous* if its automorphism group acts transitively on $\mathcal{C}$, and *regular* if it is pointed (contains no lines) and strictly convex in the sense of containing no affine subspaces of positive dimension. Vinberg established that such cones correspond bijectively to commutative T-algebras, which may be realized as matrix algebras with entries in auxiliary vector spaces V_{ij} equipped with bilinear products subject to compatibility constraints.

Within the classical theory, the irreducible homogeneous cones fall into four well-known types, arising from matrix cones over the division algebras $\mathbb{R}$, $\mathbb{C}$, $\mathbb{H}$, and $\mathbb{O}$. Beyond these, a "fifth class" appears and, quite strikingly, within non-self-dual cones associated to probability theory and information geometry.

© The Author(s), under exclusive license to Springer Nature Switzerland AG 2026

N. C. Combe et al., *Exploring Information Geometry*, Frontiers in Mathematics,
https://doi.org/10.1007/978-3-032-23054-6_12

A Jordan algebra is a commutative algebra $(J, \bullet)$ satisfying the Jordan identity

$$(x^2 \bullet y) \bullet x = x^2 \bullet (y \bullet x),$$

where $x^2 := x \bullet x$. The algebra J is called *formally real* if $x_1^2 + \cdots + x_k^2 = 0$ implies $x_i = 0$ for all i. The celebrated Koecher–Vinberg theorem [Koe69, V63] establishes a bijection between formally real Jordan algebras and *symmetric cones*, i.e., open convex cones that are both homogeneous and self-dual. Explicitly,

$$\{\text{Formally real Jordan algebras}\} \quad \longleftrightarrow \quad \{\text{Symmetric cones}\}.$$

For example, the cone of positive-definite real symmetric matrices corresponds to the Jordan algebra $\mathcal{S}_n(\mathbb{R})$ with the symmetrized product $X \bullet Y = \frac{1}{2}(XY + YX)$.

12.1 Jordan Algebras and Mirrors

For the reader's convenience recall some known results concerning Jordan algebras, Peirce's decomposition theorem and Peirce's mirror/ reflection. For more information, see [MC04, Pe81].

12.1.1 On Peirce's Reflections

Peirce's decomposition arises in the context of associative rings and also Jordan algebras (see Sect. 6.1 in [MC04]). The Peirce decomposition theorem of an associative algebra (or ring) states that one can decompose this algebra (or ring) as a sum of eigenspaces of commuting idempotent elements.

Consider an idempotent a of an algebra A. An idempotent is an element $a \in A$ which verifies $a^2 = a$. Then, there exist right, left and two-sided Peirce decompositions. These decompositions are respectively below:

$$A = Aa + A(1 - a),$$

$$A = aA + (1 - a)A,$$

$$A = aAa + aA(1 - a) + (1 - a)Aa + (1 - a)A(1 - a).$$

The latter writes A as the direct sum of aAa, $aA(1 - a)$, $(1 - a)Aa$ and $(1 - a)A(1 - a)$.

More generally, consider an associative ring. A decomposition $\sum_i a_i = 1$ of the unit into mutually *orthogonal idempotent elements*, (i.e. verifying the condition $a_i a_j = a_j a_i = 0$) produces a decomposition of the algebra into so-called Peirce spaces:

$$A = \sum_{i,j} a_i A a_j.$$

Remark 12.1 Note that analyzing the structure of individual pieces and how they reassemble to form A gives us the structure of the entire algebra.

Concerning Jordan algebras, they have similar Peirce decompositions. We call Jordan algebra a commutative algebra J where the following Jordan identity holds for all x, y in J:

$$x \bullet (x^2 \bullet y) = x^2 \bullet (x \bullet y). \tag{12.1}$$

Remark 12.2 Any real (associative) algebra $(A, +)$ defines a Jordan algebra $(J, \bullet)$, under the quasi-multiplication $x \bullet y := \frac{1}{2}(xy + yx)$, where $x, y \in A$. Indeed, the product is commutative and satisfies the identity of Eq. 12.1.

The only difference with the case above is that the "off-diagonal" Peirce spaces J_{ij}, where $(i \neq j)$ behave like the sum of two associative Peirce spaces $A_{ij} + A_{ji}$, giving thus a symmetry $J_{ij} = J_{ji}$. For the case where $n = 2$, the decomposition is determined by a single idempotent $a = a_1$, with its complement $a' := 1 - a$, and can be described in a simpler way. In particular, this leads to the Peirce Decomposition Theorem.

Theorem 12.1 (Peirce decomposition, Sect. 6.1 *[MC04]*, pp. 99–100) *Let J be a Jordan algebra and $a \in J$ an idempotent element. The idempotent a determines a Peirce decomposition of the algebra into the direct sum of Peirce subspaces:*

$$J = J_2 \oplus J_1 \oplus J_0 \quad J_k := J_k(a) := \{x \mid V_a x = kx\}, \quad (V \textit{ is a linear space}).$$

The diagonal Peirce subalgebras: $J_2 = U_a J$, $J_0 = U_{1-a} J$ correspond to the inner ideals.

A more general statement can be formulated. A decomposition of the unit of a Jordan algebra J into a sum of n supplementary orthogonal idempotents $a_1 + \cdots + a_n = 1$ leads to a Peirce decomposition of the algebra into a direct sum of Peirce subspaces:

$$J = \oplus_{i \leq j} J_{ij},$$

where diagonal Peirce subalgebras correspond to the inner ideals.

$$J_{ii} := U_{a_i}(J) = \{x \in J \mid a_i \bullet x = x\}$$

and the "off-diagonal" Peirce spaces are

$$J_{ij} = J_{ji} := U_{a_i,a_j}(J) = \{x \in J \mid a_i \bullet x = a_j \bullet x = \frac{1}{2}x\} \quad (i \neq j).$$

The symmetry $J_{ij} = J_{ji}$ implies that multiplication rules for these subspaces are less uniform to describe. The product of two Peirce spaces vanishes unless two of the indices can be linked.

The notion of Peirce reflection (or Peirce mirror) arises very naturally. Consider J a unital Jordan algebra and let a, a' be a pair of orthogonal idempotents. Take $u = a - a'$. Then, it is easy to check that u is an involution. Now, following Lem. 6.10 in [MC04] given a unital Jordan algebra J and u an involution in J, the involution U_u is an involutionary automorphism of J, implying that $U_u^2 = 1_J$. Therefore, the involution defined as

$$U_u = \mathrm{E}_2 - \mathrm{E}_1 + \mathrm{E}_0$$

has a description as a Peirce reflection, in terms of Peirce projections $\mathrm{E}_i(a)$ (see p.238 [MC04] for more details).

We will use this Peirce decomposition construction and reflection on the algebra $\mathfrak{P}$ of paracomplex numbers and then study the consequences for the module over this algebra.

12.1.2 Properties Related to Paracomplex Algebras

As we have already mentioned, the algebra of paracomplex numbers has a remarkable incidence on the manifold of probability distributions. We demonstrate this in the following statement which is a preliminary result to be used in the realm of statistical manifolds.

Proposition 12.1 *Consider an affine, symmetric space over a Jordan algebra. There exists exactly two affine and flat connections on this space if and only if the algebra is of rank 2, and generated by* $\{1, \varepsilon\}$ *with* $\varepsilon^2 = 1$ *or* -1.

In the case where $\varepsilon^2 = -1$, we have a complex structure. Similarly, if $\varepsilon^2 = 1$ we have a paracomplex structure .

In this proof, we use Einstein's summation convention for simplicity.

Proof (1) Suppose that we are working on an affine space over a Jordan algebra A of rank two, with basis elements $\{e_1, e_2\} = \{1, \epsilon\}$. The affine representation of the algebra, or free module AE_n, admits a real interpretation in the affine space E_{2n}([Ro97], Sect. 2.1.2). In this interpretation each vector $\mathbf{x} = (x^i) \in AE_n$ with coordinates $x^i = x^{(i,\alpha)}e_\alpha$, is interpreted as the vector $\mathbf{x} = (x^{(i,\alpha)}) \in E_{2m}$.

Let us introduce a parametrizable curve $x^i = x^i(t) \in E_{2n}$, and a tangent vector $\mathbf{w}$ to it, at a given point. Our aim is to proceed to the parallel transport of this vector, along that curve in E_{2n}. We have a parallel transport of $\mathbf{w}$ along of the curve $x^i = x^i(t)$ given by:

$$d\mathbf{w} + \Gamma \mathbf{w} d\mathbf{x} = 0,$$

and Γ is an affine connection for the A-space. Because of the splitting property, we can write the parallel transport equation in the following way:

$$d(\mathbf{w}^{(1)} \oplus \mathbf{w}^{(2)}) + (\Gamma^{(1)} \oplus \Gamma^{(2)})(\mathbf{w}^{(1)} \oplus \mathbf{w}^{(2)}) d(\mathbf{x}^{(1)} \oplus \mathbf{x}^{(2)}) = 0,$$

therefore, giving us:

$$d\mathbf{w}^{(\alpha)} + \Gamma^{(\alpha)} \mathbf{w}^{(\alpha)} d\mathbf{x}^{(\alpha)} = 0, \quad \alpha \in \{1, 2\}.$$

So, we can define an affine connection in E_{2n} having two components with respect to the local coordinates $x^{(i,\alpha)}$.

(2) Consider an affine and symmetric space over a Jordan algebra A, and suppose that there are 2 flat, affine connections on this space. These flat affine connections are constructed from a field of objects, having components:

$$\Gamma^{i}_{jk} = \Gamma^{i\alpha}_{jk} e_{\alpha} \in A.$$

Suppose that $\mathbf{v}^i = \mathbf{v}^{(i,\alpha)} e_\alpha$ are quantities from the algebra corresponding to a tangent vector $\mathbf{v}$. Then, from the following condition

$$d\mathbf{v}^i + \Gamma^i_{jk} \mathbf{v}^j d\mathbf{x}^k = 0,$$

we can define an affine connection in the affine space E^{2n} having the following components

$$\Gamma^{(i,\alpha)}_{(j,\beta)(k,\gamma)} = \Gamma^{is}_{jk} C^{\delta}_{s\beta} C^{\alpha}_{\delta\gamma},$$

where the $C^{\alpha}_{\beta\gamma}$ are structure constants of algebra A, with respect to the local adapted coordinates $x^{(\alpha,i)}$. Now, these objects are indexed by the number of generators of the algebra A. Since there exist 2 connections, it implies that $s \in \{1, 2\}$ and so that the number of generators of the algebra A is 2. □

This leads us to the following salient proposition:

Proposition 12.2 *The space of* 0*-pairs within the projective space* $\mathbb{P}^n$ *is isometric to the hermitian projective space over the algebra of paracomplex numbers.*

Proof Refer to Sect. 4.4.5 of [Ro97] for detailed proof. □

12.1.3 Geometry and Algebra of ∞-Dimensional Spaces in Caph(Ω, S, Z)

We now turn our attention to the geometric structure of spaces of probability distributions/ measures in the setting of Caph($\Omega, \mathbf{S}, \mathbf{Z}$) which is rich from the perspective of projective geometry and paracomplex geometry.

Let Caph($\Omega, \mathbf{S}, \mathbf{Z}$) be defined as in Cha. 1, Sec.9.1.

> Observe that in what follows we are NOT considering the case of the simplex, nor the framework of Definition 9.4!

Let $(\Omega, \mathbf{S}, \lambda)$ be a measure space, where λ is a fixed reference measure (e.g. Lebesgue measure on $\mathbb{R}^n$). We introduce the three following spaces.

1. Let us consider the space W of signed measures (bounded variations), defined as:

$$W := \Big\{\mu \text{ signed measure of bounded variation on } (\Omega, \mathbf{S}) : \mu \ll \lambda\Big\},$$

identified with their Radon–Nikodym densities in $L^1(\Omega, \lambda)$.

2. We also introduce the following **positive cone:**

$$\mathcal{C} := \Big\{\mu \in W : \tfrac{d\mu}{d\lambda} > 0 \ \lambda\text{-a.e.}\Big\}.$$

3. The last space $\mathcal{H}$ consists of probability measures that are absolutely continuous with respect to a fixed reference measure λ and share the same σ-ideal of null sets $\mathbf{Z} = A \in \mathbf{S} : \lambda(A) = 0$. This homogeneity in null sets means all measures in $\mathcal{H}$ are mutually absolutely continuous with λ, ensuring $P(A) = 0 \iff A \in \mathbf{Z}$ for any $P \in \mathcal{H}$. More concretely, we can express $\mathcal{H}$ as follows:

$$\mathcal{H} := \Big\{\mu \in \mathcal{C} : \langle 1, \mu\rangle = 1\Big\}, \qquad \langle f, \mu\rangle := \int_{\Omega} f \, d\mu.$$

To establish the exponential transport group action on $\mathcal{H}$, consider the action of the group $\mathfrak{G} = W^*$ (the dual space of W) on the cone $\mathcal{C}$.

The dual space W^* is the space of continuous linear functionals on W. The structure of W^* depends on the measure space $(\Omega, \mathbf{S}, \lambda)$ in the sense that whenever λ is σ-finite:

$$W^* \cong L^\infty(\Omega, \mathbf{S}, \lambda).$$

Every continuous linear functional $\ell \in W^*$ corresponds to a unique $h \in L^\infty(\lambda)$ such that

$$\ell(\mu) = \int_\Omega h\, d\mu = \int_\Omega h \cdot \frac{d\mu}{d\lambda}\, d\lambda, \qquad \forall\, \mu \in W.$$

Here, h is a bounded measurable function, and $\|h\|_{L^\infty}$ defines the operator norm.[a]

Since exponentiation does not preserve total mass, we pass to the quotient

$$\widetilde{\mathfrak{G}} \;=\; W^*/\mathbb{R},$$

where two functions $h_1, h_2 \in W^*$ are identified if they differ by a constant. For $[h] \in \widetilde{\mathfrak{G}}$ and $\mu \in \mathcal{C}$, the action is

$$[h] \cdot \mu \;=\; e^h \mu.$$

Here, the representative h is only defined modulo constants, but since multiplying by e^c with $c \in \mathbb{R}$ simply rescales the measure, the action on $\mathcal{C}$ is well-defined.

When restricted to the probability measures $\mathcal{H} \subset \mathcal{C}$, normalization yields

$$[h] \cdot \mu \;=\; \frac{e^h \mu}{\int_\Omega e^h\, d\mu}, \qquad [h] \in \widetilde{\mathfrak{G}},\;\; \mu \in \mathcal{H}.$$

This ensures the action preserves probability measures.

[a] When λ is not σ-finite then W^* is larger than $L^\infty(\lambda)$ and consists of all bounded finitely additive set functions that are absolutely continuous with respect to λ. Explicitly, W^* is isomorphic to the space of bounded λ-absolutely continuous finitely additive measures on $\mathbf{S}$.

Proposition 12.3 (Banach Manifold Structure) *Let* $(\Omega, \mathbf{S}, \lambda)$ *be a* σ*-finite measure space. Then:*

1. *The space of probability measures*

$$\mathcal{H} = \Big\{ \mu \in \mathcal{C} : \langle 1, \mu \rangle = 1 \Big\}$$

is an infinite-dimensional Banach manifold.

2. *The exponential transport group* $\widetilde{\mathfrak{G}} = L^{\infty}(\lambda)/\mathbb{R}$ *acts transitively and freely on* $\mathcal{H}$ *via*

$$[h] \cdot \mu = \frac{e^h \mu}{\langle e^h, \mu \rangle}, \qquad [h] \in L^{\infty}(\lambda)/\mathbb{R}.$$

This action is well-defined, smooth, and preserves the σ*-ideal of null sets* $\mathbf{Z}$*.*

The proof is available in the literature. We do not present it here since it goes beyond the tools presented in the book.

12.1.4 Paracomplex Geometry of Caph$(\Omega, \mathrm{S}, \mathrm{Z})$ (Continuous Case)

> We now show the existence of paracomplex manifold structures within certain manifolds in the class Caph$(\Omega, \mathbf{S}, \mathbf{Z})$.

In paracomplex geometry, a *paracomplex manifold* $(\mathcal{M}, \mathfrak{K})$ is a smooth manifold $\mathcal{M}$ endowed with a smooth endomorphism

$$\mathfrak{K} \colon \mathcal{T}(\mathcal{M}) \longrightarrow \mathcal{T}(\mathcal{M})$$

of the tangent bundle satisfying the following conditions:

1. **Involutivity:**

$$\mathfrak{K}^2 = \mathrm{id}_{\mathcal{T}(\mathcal{M})}.$$

2. **Eigendistributions:** The ± 1–eigenbundles

$$D_+ := \ker(\mathfrak{K} - \mathrm{id}), \qquad D_- := \ker(\mathfrak{K} + \mathrm{id})$$

have equal rank, and each is integrable in the sense of Frobenius.

Theorem 12.2 (Paracomplex Geometry of $\mathcal{H}$) *Assume* $(\Omega, \mathbf{S})$ *admits a measurable involution*

$$\tau : \Omega \to \Omega, \qquad \tau^2 = \mathrm{id},$$

such that:

1. τ *decomposes* $\Omega = \Omega^+ \sqcup \Omega^-$ *with* $\Omega^+ \cap \Omega^- = \varnothing$,
2. $\mu(\Omega^\pm) > 0$ *for all* $\mu \in \mathcal{H}$.

Then $\mathcal{H}$ *admits a natural paracomplex structure*

$$\mathfrak{K} : \mathcal{T}(\mathcal{H}) \to \mathcal{T}(\mathcal{H}),$$

defined at $\mu \in \mathcal{H}$ *by*

$$(\mathfrak{K}h)(x) = \begin{cases} h(x), & x \in \Omega^+, \\ -h(x), & x \in \Omega^-, \end{cases} \qquad \forall\, h \in \mathcal{T}_\mu(\mathcal{H}).$$

Moreover, the Fisher metric

$$g_\mu(h, k) = \int_\Omega h\, k\, d\mu$$

is para-Kähler in the sense that

$$g_\mu(\mathfrak{K}h, \mathfrak{K}k) = -\, g_\mu(h, k).$$

Proof **Step 1. Paracomplex structure.**

Define

$$\mathfrak{K} : \mathcal{T}_\mu(\mathcal{H}) \to \mathcal{T}_\mu(\mathcal{H}), \qquad (\mathfrak{K}h)(x) = \begin{cases} h(x), & x \in \Omega_+, \\ -h(x), & x \in \Omega_-. \end{cases}$$

Then

$$(\mathfrak{K}^2 h)(x) = \mathfrak{K}(\mathfrak{K}h)(x) = \begin{cases} (\mathfrak{K}h)(x), & x \in \Omega_+, \\ -(\mathfrak{K}h)(x), & x \in \Omega_- \end{cases} = h(x).$$

Hence $\mathfrak{K}^2 = \mathrm{id}$.

The ± 1 eigendistributions are

$$D_\pm = \{h \in \mathcal{T}_\mu(\mathcal{H}) : \mathfrak{K}h = \pm h\}.$$

Explicitly, D_+ consists of h supported in Ω_+, and D_- consists of h supported in Ω_-. Since $\Omega = \Omega_+ \sqcup \Omega_-$ is a measurable partition, $T_\mu\mathcal{H} = D_+ \oplus D_-$, so $\mathfrak{K}$ is integrable.

Step 2. Para-Kähler metric.

Define a pseudo-Riemannian metric by

$$g_\mu(h,k) = \int_{\Omega_+} hk\,d\mu - \int_{\Omega_-} hk\,d\mu.$$

This has neutral signature (∞, ∞) since $L^2(\mu)$ decomposes as $L^2(\mu|_{\Omega_+}) \oplus L^2(\mu|_{\Omega_-})$, with one positive and one negative block.

Compatibility: for $h, k \in \mathcal{T}_\mu(\mathcal{H})$,

$$g_\mu(\mathfrak{K}h, \mathfrak{K}k) = \int_{\Omega_+} (\mathfrak{K}h)(\mathfrak{K}k)\,d\mu - \int_{\Omega_-} (\mathfrak{K}h)(\mathfrak{K}k)\,d\mu = g_\mu(h,k).$$

Thus $(g, \mathfrak{K})$ define a paracomplex Hermitian structure.

Define the fundamental 2-form

$$\omega_\mu(h,k) = g_\mu(\mathfrak{K}h, k).$$

Then ω_μ is skew-symmetric and nondegenerate. In the information-geometric setting, ω_μ is closed, hence $(\mathcal{H}, g, \mathfrak{K}, \omega)$ is para-Kähler. □

Example 12.1 (Gaussian measures on $\mathbb{R}$) We discuss the following example of Gaussian measures. Let $\Omega = \mathbb{R}$, where $\tau(x) = -x$, so that $\Omega_+ = [0, \infty)$ and $\Omega_- = (-\infty, 0)$. The Gaussian measure is

$$d\mu(x) = \frac{1}{\sqrt{2\pi\sigma^2}} \exp\Big(-\frac{(x-\mu)^2}{2\sigma^2}\Big)\,dx.$$

Manifold Structure. The Gaussian family is parametrized by

$$(\mu, \sigma^2) \in \mathbb{R} \times \mathbb{R}_{>0},$$

giving a 2-dimensional submanifold of $\mathcal{H}$. The tangent vectors are:

$$\partial_\mu = \frac{x-\mu}{\sigma^2}, \qquad \partial_{\sigma^2} = \frac{(x-\mu)^2}{2\sigma^4} - \frac{1}{2\sigma^2}.$$

Group Action. For an affine function $h(x) = ax + b$, the action

$$\mu \mapsto \mu_h = \frac{e^h \mu}{\langle e^h, \mu\rangle}$$

yields a new Gaussian:

$$\mu_h \sim \mathcal{N}(\mu + a\sigma^2, \sigma^2).$$

Paracomplex Structure. Define $\mathfrak{K} : \mathcal{T}(\mathcal{H}) \to \mathcal{T}(\mathcal{H})$ by

$$(\mathfrak{K}h)(x) = \begin{cases} h(x), & x \geq 0, \\ -h(x), & x < 0. \end{cases}$$

On the other side the **Fisher metric** is provided by the following formula

$$g = \frac{1}{\sigma^2} d\mu^2 + \frac{2}{\sigma^4} d(\sigma^2)^2,$$

which has Lorentzian signature (1, 1).
Neutral Metric. The adjusted pseudo-Riemannian metric is

$$g_\mu(h, k) = \int_0^\infty hk \, d\mu - \int_{-\infty}^0 hk \, d\mu.$$

Verification. For $h = \partial_\mu$, $k = \partial_{\sigma^2}$:

$$g_\mu(\partial_\mu, \partial_{\sigma^2}) = \frac{1}{\sigma^2} \left(\int_0^\infty \frac{x-\mu}{\sigma^2} \, d\mu - \int_{-\infty}^0 \frac{x-\mu}{\sigma^2} \, d\mu \right) = 0.$$

Moreover,

$$g_\mu(\mathfrak{K}\partial_\mu, \mathfrak{K}\partial_{\sigma^2}) = g_\mu(\partial_\mu, \partial_{\sigma^2}) = 0.$$

Thus,

$$g_\mu(\mathfrak{K}h, \mathfrak{K}k) = g_\mu(h, k),$$

while the fundamental form

$$\omega_\mu(h, k) = g_\mu(\mathfrak{K}h, k)$$

is symplectic.

Therefore, the Gaussian exponential family furnishes a *concrete realization of this para-Kähler structure.*

Comparison of Probability Measure Spaces: Finite vs. Continuous Setting.

Aspect	Finite Ω ($\lvert\Omega\rvert = n$)	Continuous Ω (σ-finite measure space)
Manifold $\mathcal{H}$	$(n-1)$-dimensional simplex $\Delta^{n-1} \subset \mathbb{R}^n$	Infinite-dimensional Banach manifold modeled on Orlicz space $L^\Phi(\mu)$
Tangent space $\mathcal{T}_\mu(\mathcal{H})$	$\{v \in \mathbb{R}^n : \sum_{i=1}^n v_i = 0\}$	$\left\{ h \in L^\Phi(\mu) : \int_X h d\mu = 0 \right\}$ where $\Phi(x) = \cosh x - 1$
Paracomplex structure	Not present (Fisher-Rao metric positive-definite)	Exists via involution

Theorem 12.3 (Paracomplex Dolbeault Lemma)
Any (local) potential φ on a cone of probability distributions determine the local paracomplex Dolbeault $(1, 1)$*–form*

$$\widetilde{\omega} := \partial_+\partial_-\varphi = \varepsilon\partial\overline{\partial}\varphi.$$

The potential φ is defined uniquely modulo subspace of local functions $Ker\ \partial_+\partial_-$.

Proof The proof uses an explicit construction of the paracomplex structure encoded in the direct sum $\mathbb{R} \oplus \mathbb{R}'$.

Let (V, I, g) be a para–Kähler manifold with para–Kähler form ω. Consider a point p on V, and an an open neighborhood U of p. Let $(z^i_\pm)$ be adapted coordinates defined on U and mapping U onto the product of two simply connected open sets $U^\pm \subset \mathbb{R}^n$, where $n = dim_{\mathfrak{C}} V$. Moreover, assume that $z^i_\pm(p) = 0$.

Suppose that $\partial_+\theta^+ = 0$ on $U \cong U^+ \times U^-$. The Dolbeault paracomplex technique shows that there exists a function φ^+ on U, given by:

$$\varphi^+ := \int\limits_{(0,z_-)}^{(z_+,z_-)} \theta^+.$$

The integration is over any path from $(0, z_-)$ to (z_+, z_-) contained in $U^+ \times \{z_-\}$.

From the condition that $\partial_+\theta^+ = 0$, it follows that the integral is path independent and that the one-form θ^+ restricted to $U^+ \times \{z_-\}$ is closed (and thus exact), since U^+ is simply connected.

We now show that there exists a real valued function φ defined in some simply connected open neighborhood U of p such that $\omega = \partial_-\partial_+\varphi$ on U. The function φ is unique up to addition of a real–valued function f satisfying the equation $\partial_-\partial_+ f = 0$. Any such function is of the form $f = f_+ + f_-$, where $f_\pm : U \to \mathbb{R}$ satisfying the equation $\partial_\mp\partial_\pm f = 0$.

The first cohomology of U vanishes, so $H^1(U, \mathbb{R}) = 0$. Since ω is closed, there exists a one–form θ such that $\omega = d\theta$. We decompose θ into its homogeneous components: $\theta = \theta^+ + \theta^-, \theta^+ \in \Omega^{1,0}(U), \theta^- \in \Omega^{0,1}(U)$. Then

$$d\theta = \partial_+\theta^+ + (\partial_-\theta^+ + \partial_+\theta^-) + \partial_-\theta -.$$

From the fact that ω is of the type (1,1), we obtain the equations:

$$\partial_\pm\theta^\pm = 0, \quad \text{and} \quad \partial_-\theta^+ + \partial_+\theta^- = \omega.$$

Therefore, there exist two real-valued functions $\varphi^\pm$ such that $\partial_\pm\varphi^\pm = \theta^\pm$. Assuming that $\varphi := \varphi^+ - \varphi^-$, we have:

$$\partial_-\partial_+\varphi = \partial_-\partial_+\varphi^+ + \partial_+\partial_-\varphi^- = \partial_-\theta^+ + \partial_+\theta^- = \omega.$$

It is clear that the function φ is unique up to adding a solution of $\partial_-\partial_+ f = 0$ (in fact, any solution is of the form $f = f_+ + f_-$, where $\partial_\pm f_\pm = 0$.). Let us consider $\partial_+ f = \sum f_i^+ dz_+^i$, with $f_i^+ = \frac{\partial f}{\partial z_+^i}$, We get

$$0 = \partial_-\partial_+ f = \sum \frac{\partial f_i^+}{\partial z_-^j} dz_-^j \wedge dz_+^i.$$

Therefore, $\frac{\partial f_i^+}{\partial z_-^j} = 0$ and the functions ∂f_i^+ depend only on the positive coordinates z_+. So, we obtain

$$f = \sum \int_0^{z_+} f_i^+(\xi) d\xi^i + f_-(0, z_-),$$

where $\xi = (\xi^1, \ldots, \xi^n)$. The path integral is well defined, U^+ being simply connected. By a change of notation, we have that $f_+ = \sum \int_0^{z_+} f_i^+(\xi) d\xi^i$, and thus $f = f_+ + f_-$.

Conversely, let φ be a real-valued function on $U \subset V$ such that $\omega = \partial_-\partial_+\varphi$ is a non-degenerate two–form. This two–form is closed and of type (1,1). This is equivalent to $I^*\omega = -\omega$, which implies that $g := \omega(I\cdot, \cdot)$ is symmetric i.e. $g(X, Y) = g(Y, X)$. □

12.1.5 Para-Kähler geometry of exponential families

We consider a d-dimensional exponential family with natural parameter space $\Theta \subset \mathbb{R}^d$. Let $\mathcal{M}$ be an exponential family regarded as a dually flat manifold $(\mathcal{M}, g, \nabla^{(e)}, \nabla^{(m)})$, where g is the Fisher information metric, $\nabla^{(e)}$ is the exponential connection, and $\nabla^{(m)}$ the mixture connection. We denote by $\Theta = (\theta^1, \ldots, \theta^d)$ the natural (exponential) coordinates, and by $\eta = \nabla\psi(\theta)$ the expectation (mixture) coordinates associated to the cumulant generating function ψ.

On the cotangent bundle $\mathcal{T}^*(\mathcal{M})$ we take local coordinates

$$(\theta^1, \ldots, \theta^d, \eta_1, \ldots, \eta_d), \qquad \eta_i = \frac{\partial \psi}{\partial \theta^i}.$$

Here, θ represent the exponential parameters while η are the corresponding mean parameters. The tangent space at a point has a basis:

$$\{\frac{\partial}{\partial \theta^1}, \ldots, \frac{\partial}{\partial \theta^d}, \frac{\partial}{\partial \eta^1}, \ldots, \frac{\partial}{\partial \eta^d}.\}$$

The duality relation ensures independence on $\mathcal{T}^*(\mathcal{M})$.

Define the endomorphism $\mathfrak{K} : \mathcal{T}(\mathcal{T}^*(\mathcal{M})) \to \mathcal{T}(\mathcal{T}^*(\mathcal{M}))$ on the tangent bundle of the cotangent bundle of a dually flat manifold by

$$\mathfrak{K}\left(\frac{\partial}{\partial\theta^i}\right) = \frac{\partial}{\partial\theta^i}, \qquad \mathfrak{K}\left(\frac{\partial}{\partial\eta_j}\right) = -\frac{\partial}{\partial\eta_j}.$$

Then $\mathfrak{K}^2 = \mathrm{id}$ and

$$\mathcal{T}(\mathcal{T}^*(\mathcal{M})) = \mathcal{L}^+ \oplus \mathcal{L}^-, \qquad \operatorname{rank} \mathcal{L}^\pm = d,$$

where $\mathcal{L}^\pm$ are the ± 1 eigenbundles. Geometrically, $\mathcal{L}^+$ corresponds to exponential directions (θ-variations) and $\mathcal{L}^-$ to mixture directions (η-variations).

The Fisher metric g extends naturally to a neutral metric on $\mathcal{T}^*(\mathcal{M})$:

$$g = \sum_{i=1}^{d} \left(d\theta^i \otimes d\eta_i + d\eta_i \otimes d\theta^i\right).$$

This metric has signature (d, d) and satisfies

$$g(\mathfrak{K}X, \mathfrak{K}Y) = -\, g(X, Y).$$

The canonical symplectic structure is

$$\Omega = \sum_{i=1}^{d} d\theta^i \wedge d\eta_i,$$

which relates to g and $\mathfrak{K}$ via

$$\Omega(X, Y) = g(\mathfrak{K}X, Y).$$

Thus $(g, \mathfrak{K}, \Omega)$ forms a para-Kähler triple naturally associated to the dualistic geometry of exponential families.

We provide an interpretation.

Object	Exponential foliation (θ)	Mixture foliation (η)
Base vectors	$d\theta^i$ (natural parameters)	$d\eta_j$ (expectation parameters)
Leaves	$\eta =$ const (e-flat)	$\theta =$ const (m-flat)
Curvature	$\nabla^{(e)}$-flat	$\nabla^{(m)}$-flat
$\mathfrak{K}$-action	$+1$ eigenvectors	-1 eigenvectors

12.2 Frobenius Structures on Submanifolds of $\mathcal{H}$

It has been proven that certain spaces of probability measures of exponential type carry the structure of a Frobenius manifold [CM20]. The space of probability measures $\mathcal{H}$ mentioned previously (with the defined structure) does not admit a Frobenius manifold structure in the infinite-dimensional case as such. However, finite-dimensional submanifolds of $\mathcal{H}$ can be Frobenius manifolds under specific conditions. Below, we characterize these cases with rigorous proofs and examples.

A finite-dimensional exponential submanifold $\mathcal{M} \subset \mathcal{H}$ admits a Frobenius structure if and only if:

1. $\mathcal{M}$ is **flat** under the Fisher metric (i.e., $\nabla^2\psi$ induces a flat Levi-Civita connection).
2. The cumulant function ψ is a **cubic polynomial** in flat coordinates.
3. ψ satisfies the **WDVV equations** (and note that this is automatic if ψ is cubic in flat coordinates).

Frobenius structures exist for example for specific flat 1D families with quadratic ψ.The infinite-dimensional space $\mathcal{H}$ admits no Frobenius structure, if we use the standard WDVV setup mentioned in this manuscript.

Non-Examples and Why They Fail

- **Full Space of densities:** The infinite-dimensional case admits non-flat curvature. For $\mathcal{H}$ on $\mathbb{R}$, the Fisher metric has for instance non-zero curvature tensor.
- **Exponential Distribution:** $\mathcal{M} = \{p(x; \lambda) = \lambda e^{-\lambda x} \mid \lambda > 0\}$. 1-dimensional; no non-trivial Frobenius structure exists in 1D.
- **Poisson Family:** $\mathcal{M} = \{p(k; \lambda) = e^{-\lambda}\frac{\lambda^k}{k!} \mid \lambda > 0\}$. Metric $g = \lambda^{-1}d\lambda^2$ has no cubic potential satisfying WDVV.

Proposition 12.4 *A finite-dimensional exponential family $\mathcal{M} \subset \mathcal{H}$ being locally isomorphic to a product of one-dimensional flat exponential families, where each factor is one of the following:*

- *Gaussian with fixed mean (variance varies);*
- *Gaussian with fixed variance (mean varies);*
- *Exponential distribution;*
- *Gamma distribution with fixed shape (scale varies).*

admits the structure of a Frobenius manifold. In all cases, the cumulant function ψ is quadratic in flat coordinates (i.e., a degenerate cubic).

Proof Assume that $\mathcal{M}$ is a product of k one-dimensional flat exponential families of the types listed above. We prove that $\mathcal{M}$ admits a Frobenius structure.

Each one-dimensional factor has a flat Fisher metric in suitable coordinates:

- Gaussian (fixed mean): flat in $u = \log \sigma^2$,
- Gaussian (fixed variance): flat in μ,
- Exponential: flat in $t = \mathbb{E}[x] = 1/\theta$,
- Gamma (fixed shape): flat in $t = \mathbb{E}[x] = k/\theta$.

For the product manifold $\mathcal{M} = \mathcal{M}_1 \times \cdots \times \mathcal{M}_k$, the Fisher metric is the direct sum of the factor metrics,

$$g = g_1 \oplus \cdots \oplus g_k.$$

Since each g_i is flat, g is flat, and the Levi-Civita connection satisfies $R = 0$.

In flat coordinates $(t_1, \dots, t_d)$ for $\mathcal{M}$, the metric is constant (diagonalized as $g_{ij} = \delta_{ij}$ after rescaling). Each factor's cumulant function ψ is quadratic in flat coordinates, for instance:

$$\psi(u) = \frac{1}{4}u^2 \text{ (Gaussian with fixed mean)}, \qquad \psi(t) = -\log t \text{ (Exponential)},$$

with Legendre dual $\Phi(t) = \frac{1}{2}t^2$. For the product, define the Frobenius potential

$$\Phi(t) = \sum_{i=1}^{d} c_i t_i^2, \quad c_i \in \mathbb{R}.$$

This is a degenerate cubic, since any cubic term with distinct indices vanishes. Then $g_{ij} = \partial_i \partial_j \Phi$ and $\partial_i \partial_j \partial_k \Phi = 0$ for all i, j, k.

The structure constants $c_{ijk} = \partial_i \partial_j \partial_k \Phi = 0$ imply the product on $\mathcal{T}_p(\mathcal{M})$ is trivial:

$$\partial_{t_i} \circ \partial_{t_j} = 0.$$

This product is commutative and associative. The usual unit axiom $e * X = X$ is relaxed in the degenerate case; define $e = \sum_i \partial_{t_i}$. Metric invariance holds trivially,

$$g(X \circ Y, Z) = g(X, Y \circ Z) = 0,$$

and the WDVV equations hold vacuously as $0 = 0$.

Explicitly, the WDVV equations

$$\sum_{p,q} \partial_i \partial_j \partial_p \Phi \; g_{pq} \; \partial_q \partial_k \partial_l \Phi = \sum_{p,q} \partial_i \partial_k \partial_p \Phi \; g_{pq} \; \partial_q \partial_j \partial_l \Phi$$

are satisfied identically since all third derivatives vanish. □

12.2.1 Conclusion on Paracomplex and Frobenius Constructions

We now return to the analysis of a case where the manifold in question is an F-manifold–a weaker variant of a Frobenius manifold that does not require a metric–endowed with a paracomplex structure, a situation that naturally arises in the statistical setting.

Theorem 12.4 *([CCN22]) The paracomplex Frobenius manifold $\mathcal{M}$ has the following properties:*

1. *$\mathcal{M}$ is identified with a projective manifold defined over the rank-two paracomplex algebra $\mathfrak{P}$.*
2. *$\mathcal{M}$ decomposes into a pair of isomorphic totally geodesic submanifolds $\mathcal{M}_+$ and $\mathcal{M}_-$, equipped with flat connections ∇ and ∇^* respectively.*
3. *The submanifolds $\mathcal{M}_+$ and $\mathcal{M}_-$ are symmetric with respect to a Peirce reflection.*

Proof **Part 1: Projective structure over $\mathfrak{P}$**
Let $\mathfrak{P} = \mathbb{R} \oplus \mathbb{R}$ be the rank-two paracomplex algebra with product $(a, b) \cdot (c, d) = (ac, bd)$ and involution $(a, b)^\dagger = (b, a)$[1]. The projective space $\mathbb{P}^1(\mathfrak{P})$ consists of equivalence classes $[z : w]$ where $(z, w) \in \mathfrak{P}^2 \setminus \{(0, 0)\}$ with identification:

$$(z, w) \sim (\lambda z, \lambda w), \quad \lambda \in \mathfrak{P}^\times.$$

The Frobenius potential $\Phi : \mathfrak{P}^2 \to \mathbb{R}$ is defined by:

$$\Phi(z, w) = \frac{1}{2} \operatorname{Re} \left((z^\dagger w)^2 \right)$$

where $\operatorname{Re}(a, b) = a + b$. This potential is $\mathfrak{P}^\times$-homogeneous:

$$\Phi(\lambda z, \lambda w) = |\lambda|^4 \Phi(z, w), \quad |\lambda|^2 = \lambda^\dagger \lambda.$$

[1] The symbol † denotes the swap involution on the paracomplex algebra $\mathfrak{P} = \mathbb{R} \oplus \mathbb{R}$. Specifically, for any element $(a, b) \in \mathfrak{P}$,

$$(a, b)^\dagger = (b, a).$$

.

The manifold $\mathcal{M} = \mathbb{P}^1(\mathfrak{P})$ is embedded via:

$$[z : w] \mapsto \big(\mathrm{Re}(z^\dagger w), \mathrm{Im}(z^\dagger w)\big) \in \mathbb{R}^2$$

with $\mathrm{Im}(a, b) = a - b$. The Frobenius structure is induced by the Hessian $\nabla^2 \Phi$.

Part 2: Totally geodesic decomposition

Consider the idempotents $e_1 = (1, 0), e_2 = (0, 1) \in \mathfrak{P}$. Define submanifolds:

$$\mathcal{M}_+ = \{[z : 0] \mid z \in \mathfrak{P} \setminus \{0\}\} \cong \mathbb{P}^1(\mathbb{R}) \cong S^1,$$
$$\mathcal{M}_- = \{[0 : w] \mid w \in \mathfrak{P} \setminus \{0\}\} \cong \mathbb{P}^1(\mathbb{R}) \cong S^1.$$

Each is equipped with connections:

- On $\mathcal{M}_+$: Levi-Civita connection ∇ for metric $g_+ = dz \otimes dz$;
- On $\mathcal{M}_-$: Dual connection ∇^* for metric $g_- = -dw \otimes dw$.

Both connections are flat since g_+ and g_- are constant. Geodesics in $\mathcal{M}_+$ are curves $\gamma(t) = [z_0 + ta : 0]$ with $\nabla_{\dot\gamma}\dot\gamma = 0$. These remain geodesics in $\mathcal{M}$ as the second fundamental form vanishes:

$$\mathrm{II}(\dot\gamma, \dot\gamma) = \nabla^{\mathcal{M}}_{\dot\gamma}\dot\gamma - \nabla^{\mathcal{M}_+}_{\dot\gamma}\dot\gamma = 0.$$

Part 3: Peirce reflection symmetry

The Peirce reflection $\sigma : \mathfrak{P} \to \mathfrak{P}$ is the involution:

$$\sigma(a, b) = (b, a).$$

This lifts to a diffeomorphism $\Sigma : \mathbb{P}^1(\mathfrak{P}) \to \mathbb{P}^1(\mathfrak{P})$:

$$\Sigma([z : w]) = [w : z].$$

Σ swaps the submanifolds:

$$\Sigma(\mathcal{M}_+) = \mathcal{M}_-, \quad \Sigma(\mathcal{M}_-) = \mathcal{M}_+,$$

and is an isometry:

$$g(\Sigma_* X, \Sigma_* Y) = -g(X, Y), \quad \forall X, Y \in T\mathcal{M}.$$

The potential is invariant under Σ:

$$\Phi(\Sigma(z, w)) = \Phi(w, z) = \frac{1}{2}\mathrm{Re}((w^\dagger z)^2) = \frac{1}{2}\mathrm{Re}((z^\dagger w)^2) = \Phi(z, w).$$

Thus, $\mathcal{M}_+$ and $\mathcal{M}_-$ are symmetric with respect to Σ. □

Table 12.1 Algebraic and geometric dictionary

	Algebraic properties	Geometric interpretation
Algebra $\mathfrak{P}$	Generated by 1 and ε where $\varepsilon^2 = 1$	linear space over $\mathfrak{P}$/manifold over
Idempotents	$\mathfrak{P}$ has a pair of orthogonal idempotents: $e_- = \frac{1-\varepsilon}{2}$, $e_+ = \frac{1+\varepsilon}{2}$	Real interpretation in the real linear space: pair of m-eigenspaces E^m_+ and E^m_-
$\mathfrak{P}$-module		Tangent space T_θ at point P_θ is identified to a $\mathfrak{P}$-module
Real interpretation in the real linear space	Pair of eigenspaces $E^m_+ = \{v \in E^{2m}_{\mathfrak{K}} \mid \mathfrak{K}v = \varepsilon v\}$ $E^m_- = \{v \in E^{2m}_{\mathfrak{K}} \mid \mathfrak{K}v = -\varepsilon v\}$	$\mathfrak{P}$-manifold decomposes into a direct sum of submanifolds being totally geodesic
Connections		The submanifolds have flat connections
Peirce decomposition	Peirce decomposition and involution	The submanifolds are symmetric to each other with respect to Peirce mirror

Note that the second property corresponds to the realizations of the module over the maximal ideal of $\mathfrak{K}$ and its annihilator.

Corollary 12.1 *The results are summarized in Table 12.1.*

Comment on the Method of proof

While Norden's seminal works [No52, No54, No58] laid the groundwork for almost complex manifolds, they do not directly extend to the paracomplex (split-complex) setting required here. This extension is provided by Shirokov [Sh54, Sh02], whose formulation applies to modules over arbitrary finite, unital, commutative algebras—including those admitting nontrivial idempotents.

Within this setting, Peirce's decomposition theorem naturally enters the picture, and the presence of idempotents prompts the use of the corresponding Peirce reflections (cf.§6.1 of [MC04]). This algebraic apparatus not only adapts the idea of "complex" coordinates to the paracomplex case, but also offers an efficient tool for analyzing integrability and curvature conditions on our statistical F-manifold.

Novel Perspectives and Openings 13

We end this manuscript with a synthesis of ideas from symplectic geometry, quantum cohomology, and information geometry based on the works [CM20, CCN21].

Extending an analog of classical Gromov–Witten invariants to the realm of information geometry, we introduce Gromov–Witten invariants for statistical manifolds (denoted GWS). Originally, these invariants are rational numbers that enumerate pseudo-holomorphic curves satisfying specific conditions in a symplectic manifold. In our generalization, GWS encode fundamental geometric structures of statistical manifolds, reflecting the intersection theory of (para-)holomorphic curves within this framework.

Furthermore, this perspective reveals an intrinsic connection between the geometry of statistical learning and the dynamics of the learning process. The presence or obstruction of certain pseudo-holomorphic structures, as captured by GWS, provides a criterion for determining whether a learning system successfully acquires information or encounters fundamental limitations, thereby offering a novel geometric approach to the theory of learning.

13.1 Gromov–Witten Invariants: Classical and Statistical

Gromov–Witten invariants are fundamental numerical invariants in symplectic geometry and algebraic geometry, capturing intersection properties of (pseudo-)holomorphic curves in a given space [G85, KonMa94, RuTi95, W91]. Initially they emerged from string theory. In the classical setting, given a compact symplectic manifold $(\mathcal{M}, \varpi)$, the Gromov–Witten invariant counts the number of (pseudo-)holomorphic maps

$$u : (\mathscr{S}, j) \to (\mathcal{M}, \varpi),$$

© The Author(s), under exclusive license to Springer Nature Switzerland AG 2026

N. C. Combe et al., *Exploring Information Geometry*, Frontiers in Mathematics,
https://doi.org/10.1007/978-3-032-23054-6_13

from a compact Riemann surface $\mathscr{S}$ (with complex structure j) into $\mathcal{M}$, satisfying certain constraints on their homology class and intersection conditions with given cycles.

This notion appears in **Enumerative Geometry**: Gromov–Witten invariants count the number of holomorphic curves passing through prescribed points or satisfying intersection constraints. In particular, they play an important role within the context of **Quantum Cohomology,** central in the theory of formal Frobenius manifolds. The latter define a deformation of classical cohomology, giving rise to a quantum product that encodes curve counts in a ring structure. We refer to [Ma99] for more information on this topic.

The point of view adopted here is naturally inspired by *quantum cohomology*. Quantum cohomology endows the cohomology ring of a symplectic manifold with a Frobenius manifold structure where, roughly speaking, the deformation is governed by Gromov–Witten invariants, and the resulting structure is a *formal Frobenius manifold* in the sense of Dubrovin.

From this perspective, the construction developed here should be understood as a *statistical analogue*, obtained by mimicking the formalism of classical Gromov–Witten invariants in an entirely new context. In particular, the free energy $\tilde{\Phi}$ plays the role of the genus-zero potential Φ from quantum cohomology, while its derivatives (cumulants) provide the statistical counterpart of the classical GW invariants. In certain cases, the manifold of statistical models thereby acquires the structure of a Frobenius-type manifold, with entropy providing the generating potential.

> Thus, just as quantum cohomology geometrically encodes enumerative invariants of holomorphic curves, statistical Gromov–Witten invariants encode intersection data of statistical hypersurfaces. This analogy is more than formal: it points to a genuine unification of ideas across information geometry, Frobenius manifold theory, and topological field theory.

13.2 Formal Frobenius Manifold

Let (H, g) be a (formal) Frobenius manifold, where H is a finite-rank free module over a field $\mathbb{K}$ of characteristic 0 ($\mathbb{C}$ or $\mathbb{R}$), and

$$g : H \otimes H \longrightarrow \mathbb{K}$$

is a symmetric non-degenerate bilinear form (the *metric*). We denote by $H^* = \mathrm{Hom}_k(H, \mathbb{K})$ the dual module of H.

The Frobenius manifold structure is governed by a *potential function*

$$\Phi \in \mathbb{K}[[t_1, \ldots, t_r]],$$

a formal power series in flat coordinates $t = (t_1, \ldots, t_r)$ on H. Locally, Φ admits the expansion

$$\Phi(t) = \sum_{n \geq 3} \frac{1}{n!} Y_n(t^{\otimes n}),$$

where each

$$Y_n : H^{\otimes n} \longrightarrow \mathbb{K}$$

is a symmetric multilinear form. These maps Y_n are interpreted as *correlation functions* of a two-dimensional topological field theory defined on (H, g), see [Ma99].

In Gromov–Witten theory, the multilinear maps Y_n correspond to intersection numbers on the moduli space of stable maps, and the Gromov–Witten invariants are precisely the coefficients of Φ when expanded in a basis of H. This formulation provides a bridge between Frobenius manifolds, quantum cohomology, and Gromov–Witten theory, showing how the potential function encodes geometric intersection theory in an algebraic and formal power series framework. The abstract correlation functions Y_n serve as the structural foundation from which the Gromov–Witten invariants emerge, linking the geometry of moduli spaces with the algebraic structure of Frobenius manifolds.

13.3 How to Link This to Statistics?

In the framework of statistical geometry, we return to the study of statistical manifolds defined in Caph, emphasizing the discrete case of the exponential family, in the following part.

Partition functions on statistical manifolds arise by direct analogy with the partition functions of statistical mechanics in physics: just as the latter encode the weighted sum over microscopic states of a physical system, the former encode the weighted sum over probabilistic states on a statistical manifold, thereby providing the generating function for cumulants and invariants.

The *partition function* $Z(\beta)$ for parameters $\beta = (\beta^0, \cdots, \beta^n) \in \mathbb{R}^{n+1}$ is given by the expansion:

$$Z(\beta) = \sum_{\omega \in \Omega} \exp\left(-\sum_{j=0}^{n} \beta^j X_j(\omega) \right), \tag{13.1}$$

where:

- The parameter β given by $\beta = (\beta^0, \ldots, \beta^n) \in \mathbb{R}^{n+1}$ provides an affine canonical parametrization.
- $X_j : \Omega \to \mathbb{R}$ are sufficient statistics with $X_0(\omega) \equiv 1$, and $X_1(\omega), \ldots, X_n(\omega)$ are linearly independent co-vectors.

- β^0 is the log-normalization parameter ensuring $\sum_\omega p_\beta(\omega) = 1$.
- The linear functionals

$$\langle \beta, X \rangle = \sum_{j=1}^{r} \beta^j X_j(\omega)$$

belong to the dual space of the statistical manifold.

The Taylor expansion of the partition function $Z(\beta)$ generates the moments of the sufficient statistics:

$$Z(\beta) = \sum_{m=0}^{\infty} \frac{(-1)^m}{m!} \sum_{j_1,\dots,j_m} \beta^{j_1} \cdots \beta^{j_m} \sum_{\omega\in\Omega} X_{j_1}(\omega) \cdots X_{j_m}(\omega).$$

The exponential family thus forms an n-dimensional embedded submanifold of the probability simplex, uniquely determined by $n+1$ points in general position through the moment-matching conditions

$$\mathbb{E}_{p_\beta}[X_j] = \eta_j, \qquad j = 1, \dots, n.$$

The Fisher information metric

$$g_{jk} = \partial_j \partial_k \ln Z(\beta)$$

endows the exponential family with a natural Riemannian structure.

The family defines an analytic n-dimensional hypersurface within the statistical manifold, which can be uniquely determined by $n+1$ points in general position.

13.3.1 Gromov–Witten Invariants for Statistical Manifolds

We introduce the notion of Gromov–Witten invariants for statistical manifolds (GWS) as follows:

Definition 13.1 Let $\mathbb{K}$ be the field of real numbers. Let $\mathcal{S}$ be the statistical manifold forming an exponential family with canonical parameters β in a neighborhood of a base point $\beta^{(0)}$. The Gromov–Witten invariants for statistical manifolds (GWS) at $\beta^{(0)}$ are defined by the family of multilinear maps:

$$\tilde{Y}_n : T_{\beta^{(0)}}(\mathcal{S})^{\otimes n} \to \mathbb{K},$$

defined by the covariant derivatives of the log-partition function. We define the statistical potential

$$\tilde{\Phi}(\beta) = \ln \sum_{\omega\in\Omega} \exp\left(-\sum_j \beta^j X_j(\omega) \right).$$

For tangent vectors $v_1, \dots, v_n \in T_{\beta^{(0)}}(\mathscr{S})$, the statistical correlators (or *Gromov–Witten–statistical invariants*) are defined as

$$\widetilde{Y}_n(v_1, \dots, v_n) = \nabla_{v_1} \cdots \nabla_{v_n} \widetilde{\Phi}\,|_{\beta=\beta^{(0)}},$$

where ∇ denotes the Levi–Civita connection of the Fisher metric on $\mathscr{S}$.

Equivalently, in local coordinates, if

$$v_k = \sum_i a_k^i \frac{\partial}{\partial \beta^i},$$

then

$$\widetilde{Y}_n(v_1, \dots, v_n) = \sum_{i_1,\dots,i_n} a_1^{i_1} \cdots a_n^{i_n} \frac{\partial^n \widetilde{\Phi}}{\partial \beta^{i_1} \dots \partial \beta^{i_n}}\bigg|_{\beta=\beta^{(0)}}.$$

Proposition 13.1 *The Gromov–Witten invariants for statistical manifolds (GWS) at $\beta^{(0)}$ are given by the coefficients of the Taylor expansion of the log-partition function $\widetilde{\Phi}$ around $\beta^{(0)}$. Consequently, the potential $\widetilde{\Phi}$ is uniquely determined by the collection of GWS invariants.*

Proof Indeed, since $\tilde{\mathbf{\Phi}}$ in formula (13.2) relies on the polylinear maps $\tilde{Y}_n \in \left(-\sum_j \beta^j X_j(\omega)\right)^{\otimes n}$, defining the (GWS), the statement follows. □

This formulation provides a geometric interpretation in which the invariants $\widetilde{Y}_n$ capture the infinitesimal intersection properties of the hypersurfaces defined by the level sets of the sufficient statistics X_j in the statistical manifold.

The GWS invariants are built from the following objects:

1. **Generating Property**:

$$\tilde{\mathbf{\Phi}}(\beta) = \sum_{n=0}^{\infty} \frac{1}{n!} \tilde{Y}_n \left((\beta - \beta^{(0)})^{\otimes n}\right)$$

   This means that $\tilde{\mathbf{\Phi}}$ is regenerated from its covariant derivatives at $\beta^{(0)}$.
2. For sufficient statistics $\{X_j\}$, define statistical hypersurfaces

$$\mathcal{Z}(c_j) = \mathcal{Z}_j = \{p_\beta \in \mathscr{S} : \mathbb{E}_\beta[X_j] = c_j\}.$$

   Each hypersurface $\mathcal{Z}_j$ is the set of all probability distributions $p_\beta \in \mathscr{S}$ that share the same expected value for a specific observable X_j.
3. $\tilde{Y}_n$ quantifies the infinitesimal intersection density of n such statistical hypersurfaces $\mathcal{Z}_{j_1}, \cdots, \mathcal{Z}_{j_n}$ near $\beta^{(0)}$. In classical Gromov–Witten theory, invariants count intersections of cycles in a moduli space.

4. **Cumulant Correspondence**. This has an implication: $\tilde{Y}_n$ equals the n-th joint cumulant of the sufficient statistics $\{X_j\}$ evaluated at $\beta^{(0)}$.
5. **Frobenius Structure**: When $\mathscr{S} \cong \prod_{i=1}^{k} \mathcal{M}_i$ decomposes into 1D exponential families (Gaussian, exponential, or gamma with fixed parameters), $(\mathscr{S}, g, \tilde{\mathbf{\Phi}})$ forms a Frobenius manifold and $\tilde{Y}_n$ satisfies the WDVV equations:

$$\sum_{\alpha,\beta} g^{\alpha\beta} \frac{\partial^3 \tilde{\mathbf{\Phi}}}{\partial t^i \partial t^j \partial t^\alpha} \frac{\partial^3 \tilde{\mathbf{\Phi}}}{\partial t^k \partial t^l \partial t^\beta} = \sum_{\alpha,\beta} g^{\alpha\beta} \frac{\partial^3 \tilde{\mathbf{\Phi}}}{\partial t^i \partial t^k \partial t^\alpha} \frac{\partial^3 \tilde{\mathbf{\Phi}}}{\partial t^j \partial t^l \partial t^\beta}.$$

This is the associativity condition for the quantum product in Frobenius manifolds. We give it the following interpretation: $\tilde{Y}_n$ counts "virtual statistical curves" (analogous to holomorphic curves in Gromov–Witten theory) with n marked points constrained to the hypersurfaces $\mathcal{Z}_j$, in the sense that each $\mathcal{Z}_j = \{p_\beta : \mathbb{E}_\beta[X_j] \text{ fixed}\}$ corresponds to a *divisor class*. Consider two hypersurfaces $\mathcal{Z}_1$ and $\mathcal{Z}_2$:

$$\mathcal{Z}_1 \cap \mathcal{Z}_2 = \left\{p_\beta : \mathbb{E}_\beta[X_1], \mathbb{E}_\beta[X_2] \text{ fixed}\right\}.$$

This set consists of all distributions satisfying both constraints simultaneously. Their intersection

$$\mathcal{Z}_1 \cap \cdots \cap \mathcal{Z}_n$$

represents the set of probability distributions satisfying n independent constraints:

$$\mathcal{Z}_1 \cap \cdots \cap \mathcal{Z}_n = \left\{p_\beta \in \mathscr{S} : \mathbb{E}_\beta[X_j] \text{ fixed for all } j = 1, \ldots, n\right\}.$$

Now, $\tilde{Y}_n(v_1, \ldots, v_n)$ counts virtual intersections of n such divisors. The WDVV equations ensure topological consistency of these counts across different strata.

13.4 Interpretation via the Log-Partition Function and GWS Invariants

To summarize, the Gromov–Witten invariants for statistical manifolds (GWS) emerge intrinsically from the potential function $\tilde{\mathbf{\Phi}}$, which is the **log-partition function** (free energy) of the statistical system:

$$\tilde{\mathbf{\Phi}}(\beta) = \ln \sum_{\omega \in \Omega} \exp\left(-\sum_j \beta^j X_j(\omega) \right). \tag{13.2}$$

This function governs the geometry of statistical families through its covariant derivatives, which define the GWS invariants $\tilde{Y}_n$. Geometrically, these invariants encode:

- **Intersection Densities** of statistical hypersurfaces $\mathcal{Z}_j = \{p_\beta \in \mathcal{S} : \mathbb{E}_\beta[X_j] = c_j\}$.
- **Virtual Counts** of solutions to n simultaneous constraints on sufficient statistics.
- **Deformation Properties** under changes of the base distribution $p_{\beta^{(0)}}$.

This establishes a certain type of equivalence:

$$\begin{array}{ccc} \text{Frobenius potential } \Phi & \longleftrightarrow & \tilde{\mathbf{\Phi}} \\ \big\downarrow \text{GW theory} & & \big\downarrow \text{Statistical geometry} \\ \text{GW invariants } \langle [\gamma_1], \ldots, [\gamma_n] \rangle_{0,n} & \leftrightarrow & \text{GWS invariants } \tilde{Y}_n(v_1, \ldots, v_n) \end{array}$$

This allows us to formulate the following statement.

The GWS invariants $\tilde{Y}_n$ intrinsically determine the log-partition function $\tilde{\mathbf{\Phi}}$ through the **generative reconstruction** where

$$\tilde{\mathbf{\Phi}}(\beta) = \sum_{n=0}^{\infty} \frac{1}{n!} \tilde{Y}_n \left((\beta - \beta^{(0)})^{\otimes n} \right)$$

where $\tilde{Y}_n = \nabla_{v_1} \cdots \nabla_{v_n} \tilde{\mathbf{\Phi}}|_{\beta^{(0)}}$ for $v_i \in T_{\beta^{(0)}}\mathcal{S}$. Furthermore, we give a geometric characterization of $\tilde{Y}_n(v_1, \ldots, v_n)$.

13.4.1 Construction of GWS

Up to this point, we have mostly conveyed the intuition behind Gromov–Witten statistics (GWS), which is to emulate the classical construction within the statistical setting. In the following, we present the rigorous formulation of this construction.

Let us now introduce the construction of Gromov–Witten statistics (GWS), which builds upon the classical theory of Gromov–Witten invariants. A clear and elegant exposition of the classical theory can be found in [Ma99].

The Moduli Space $\mathcal{M}_{0,n}(\mathcal{S})$

The space $\mathcal{M}_{0,n}(\mathcal{S})$ denotes the moduli space of genus-zero stable maps into the statistical manifold $\mathcal{S}$ with n marked points. Formally, an element of this moduli space is a triple

$$(C; x_1, \ldots, x_n; f)$$

where:

- C is a genus-zero curve (possibly nodal).
- $x_1, \ldots, x_n$ are marked points on C.

- $f : C \to \mathscr{S}$ is a stable map (a "statistical curve").

A map is stable if it has only finitely many automorphisms. This ensures the moduli space is well-behaved (Deligne–Mumford compactification). In particular, if a component of C is contracted to a point under f, then that component must have at least three special points (marked points or nodes).

Evaluation Maps

For each marked point x_k, there is an evaluation map

$$\mathrm{ev}_k : \mathcal{M}_{0,n}(\mathscr{S}) \to \mathscr{S}, \quad \mathrm{ev}_k(C; x_1, \ldots, x_n; f) = f(x_k).$$

This map sends a stable map to the image of the k-th marked point in the statistical manifold $\mathscr{S}$.

We discuss now the constraint hypersurfaces and their classes.

The hypersurface $\mathcal{Z}_{j_k} \subset \mathscr{S}$ is defined by the moment constraint

$$\mathcal{Z}_{j_k} = \{p_\beta \in \mathscr{S} \mid \mathbb{E}_\beta[X_{j_k}] = c_{j_k}\}.$$

Its cohomology class $[\mathcal{Z}_{j_k}] \in H^*(\mathscr{S})$ formally represents this geometric condition in intersection theory.

Pullbacks and Loci of Constraints

The pullback via the evaluation map,

$$\mathrm{ev}_k^*[\mathcal{Z}_{j_k}] \in H^*(\mathcal{M}_{0,n}(\mathscr{S})),$$

is the cohomology class of the locus where the k-th marked point lies on $\mathcal{Z}_{j_k}$:

$$\{(C; x_1, \ldots, x_n; f) \in \mathcal{M}_{0,n}(\mathscr{S}) \mid f(x_k) \in \mathcal{Z}_{j_k}\}.$$

Statistically, this means the distribution at x_k satisfies $\mathbb{E}_\beta[X_{j_k}] = c_{j_k}$.

ψ-classes

The psi-class $\psi_k \in H^2(\mathcal{M}_{0,n}(\mathscr{S}))$ is the first Chern class of the cotangent line bundle at the k-th marked point. Its power $\psi_k^{d_k}$ imposes derivative conditions:

$$\begin{cases} d_k = 0 & \text{: No derivative condition (simple incidence),} \\ d_k = 1 & \text{: First-order condition (tangency),} \\ d_k = 2 & \text{: Second-order condition, etc.} \end{cases}$$

The exponent d_k quantifies the order of the derivative constraint at the k-th marked point.

The psi-classes ψ_k are therefore tautological classes on $\mathcal{M}_{0,n}$ and weight the count by the complexity of the curve at the marked point x_k, for example through derivatives of the constraints.

The Gromov–Witten Invariant

The integral

$$\int_{\mathcal{M}_{0,n}(\mathscr{S})} \prod_{k=1}^{n} \psi_k^{d_k} \cup \mathrm{ev}_k^*[\mathcal{Z}_{j_k}]$$

is a descendant Gromov–Witten invariant. It provides a virtual count of genus-zero statistical curves satisfying:

- the curve meets each constraint $\mathcal{Z}_{j_k}$ at the k-th marked point;
- with derivative conditions of order d_k at each point.

This count is "virtual" because it is defined via intersection theory on the moduli space, which may be singular.

We provide a statistical interpretation.

The quantity

$$\widetilde{Y}_n(v_1, \ldots, v_n) = \int_{\mathcal{M}_{0,n}(\mathscr{S})} \prod_{k=1}^{n} \psi_k^{d_k} \cup \mathrm{ev}_k^*[\mathcal{Z}_{j_k}]$$

represents the virtual number of solutions to the system

$$\mathbb{E}_\beta[X_{j_k}] = c_{j_k}, \quad k = 1, \ldots, n,$$

near a reference parameter $\beta^{(0)}$, counting each solution with a weight determined by the local behavior (derivatives) of the statistical curves at the constraints.

Consequently, the above discussion can be succinctly captured in the following statement.

Proposition 13.2 *The $\tilde{Y}_n$ quantify intersection densities:*

$$\tilde{Y}_n(v_1, \ldots, v_n) = \int_{\mathcal{M}_{0,n}(\mathscr{S})} \underbrace{\prod_{k=1}^{n} \psi_k^{d_k} \cup \mathrm{ev}_k^*[\mathcal{Z}_{j_k}]}_{\substack{\text{virtual number of solutions} \\ \text{to } \mathbb{E}_\beta[X_{j_k}]=c_{j_k} \\ \text{near } \beta^{(0)}}},$$

where $\mathcal{M}_{0,n}(\mathscr{S})$ is the moduli space of genus-zero stable maps into the statistical manifold $\mathscr{S}$ with n marked points. The evaluation map at the k-th marked point is defined as

$$\mathrm{ev}_k : \mathcal{M}_{0,n}(\mathscr{S}) \to \mathscr{S}, \qquad \mathrm{ev}_k\big((C; x_1, \ldots, x_n; f)\big) = f(x_k),$$

where C is a genus-zero curve (possibly nodal); $x_1, \ldots, x_n$ are marked points on C and $f : C \to \mathscr{S}$ is a "statistical curve" (a map into the manifold).

The pullback of the cohomology class of a hypersurface $\mathcal{Z}_{j_k}$ via ev_k is

$$\mathrm{ev}_k^*\big[\mathcal{Z}_{j_k}\big],$$

which represents the locus in $\mathcal{M}_{0,n}(\mathscr{S})$ where the k-th marked point lies on the hypersurface $\mathcal{Z}_{j_k}$ and $\psi_k^{d_k}$ is defined as above.

Remark 13.1 This parallels GW invariants $\langle \tau_{d_1}(\gamma_1) \cdots \tau_{d_n}(\gamma_n) \rangle_{0,n}$.

Examples for d_k.

If $d_k = 0$ for all k, the integral counts curves passing through $\mathcal{Z}_{j_k}$ with no derivative conditions.

If $d_k = 1$, the integral counts curves tangent to $\mathcal{Z}_{j_k}$ at the marked point x_k.

Higher d_k enforce higher-order tangency, for instance, matching derivatives up to order d_k.

Example: Product of Gaussian Families with Fixed Variance

Consider a statistical manifold $\mathscr{S}$ formed as the product of three independent one-dimensional Gaussian families with fixed variance $\sigma^2 = 1$. The probability density of a single Gaussian is

$$p(x; \mu) = \frac{1}{\sqrt{2\pi}} \exp\Big(-\frac{(x-\mu)^2}{2}\Big),$$

with natural parameter $\beta = \mu$ (since $\beta = \mu/\sigma^2 = \mu$). For the product family, the joint density for $\omega = (x_1, x_2, x_3)$ is

$$p(\omega; \mu_1, \mu_2, \mu_3) = \prod_{j=1}^{3} \frac{1}{\sqrt{2\pi}} \exp\Big(-\frac{(x_j-\mu_j)^2}{2}\Big).$$

The sufficient statistics are $X_j(\omega) = x_j$ for $j = 1, 2, 3$, and the natural parameters are $\beta_j = \mu_j$. Thus, the statistical manifold is $\mathscr{S} = \mathbb{R}^3$, with parameters $\beta = (\beta_1, \beta_2, \beta_3)$.

Constraints and Hypersurfaces

Impose the three moment constraints

$$\mathbb{E}_\beta[X_j] = c_j, \quad j = 1, 2, 3.$$

Since $\mathbb{E}_\beta[X_j] = \beta_j$, these define hypersurfaces

$$\mathcal{Z}_j = \{\beta \in \mathbb{R}^3 : \beta_j = c_j\},$$

which are planes in $\mathbb{R}^3$. Their intersection is the unique point

$$\beta^{(0)} = (c_1, c_2, c_3).$$

Moduli Space of Stable Maps

We consider genus-zero stable maps into $\mathcal{S} = \mathbb{R}^3$. Since $\mathbb{R}^3$ is contractible and flat (with the Fisher metric $ds^2 = \sum_{j=1}^{3} d\mu_j^2$), any holomorphic map from a genus-zero curve into $\mathbb{R}^3$ must be constant. Hence, the moduli space of stable maps is

$$\mathcal{M}_{0,3}(\mathbb{R}^3) \simeq \mathcal{M}_{0,3} \times \mathbb{R}^3,$$

where $\mathcal{M}_{0,3}$ is the moduli space of genus-zero curves with three marked points (a single point).

A constant map $f : C \to \mathbb{R}^3$ is determined by its value $\beta^0 \in \mathbb{R}^3$, independent of C. The evaluation maps satisfy

$$\mathrm{ev}_k((C; x_1, x_2, x_3; f)) = f(x_k) = \beta^0.$$

Imposing the Constraints

Pullbacks $\mathrm{ev}_k^*[\mathcal{Z}_{j_k}]$ restrict the moduli space to maps with $\beta^0 \in \mathcal{Z}_{j_k}$. For $j_k = k$, i.e., constraints X_1, X_2, X_3, we require

$$\beta^0 \in \mathcal{Z}_1 \cap \mathcal{Z}_2 \cap \mathcal{Z}_3 = \{(c_1, c_2, c_3)\}.$$

Hence, the only allowed constant map is $f(x) \equiv (c_1, c_2, c_3)$. The moduli space of maps satisfying the constraints is

$$\mathcal{M} = \mathcal{M}_{0,3} \times \{(c_1, c_2, c_3)\} \simeq \mathcal{M}_{0,3},$$

which is a single point since $\dim \mathcal{M}_{0,3} = 0$.

Psi-Classes and the Integral

On $\mathcal{M}_{0,3}$ the cotangent spaces are trivial, so the psi-classes ψ_k satisfy

$$\psi_k^{d_k} = \begin{cases} 1, & \text{if } d_k = 0, \\ 0, & \text{if } d_k > 0. \end{cases}$$

The Gromov–Witten invariant becomes

$$\widetilde{Y}_3(v_1, v_2, v_3) = \int_{\mathcal{M}_{0,3}(\mathscr{S})} \prod_{k=1}^{3} \psi_k^{d_k} \cup \mathrm{ev}_k^*[\mathcal{Z}_{j_k}] = \begin{cases} 1, & \text{if } d_1 = d_2 = d_3 = 0, \\ 0, & \text{otherwise.} \end{cases}$$

Interpretation
This result reflects that the system

$$\mathbb{E}_\beta[X_j] = c_j, \quad j = 1, 2, 3$$

has exactly one solution $\beta = (c_1, c_2, c_3)$ in $\mathbb{R}^3$. The vanishing of the invariant for $d_k > 0$ indicates that higher-order derivative conditions are unnecessary–the solution is isolated and non-degenerate.

This example illustrates how Gromov–Witten invariants encode the number of solutions to moment constraints in a flat statistical manifold, with psi-classes capturing higher-order local behavior.

13.5 Learning Geometrically

We regard the bundle $(\mathcal{T}(\mathscr{S}), \mathscr{S}, \pi)$ as the stage on which statistical dynamics unfolds.

- The **base** space is simply the statistical manifold $\mathscr{S}$ of probability distributions.
- The **fiber at** ρ is the space of centered distributions $T_\rho(\mathscr{S}) \cong \left\{ f \in L^2(\rho) : \int_\Omega f \, d\rho = 0 \right\}$,

  which correspond to score functions.
- The **structure group** is a Lie group $G \subseteq \mathrm{GL}(T_\rho(\mathscr{S}))$ acting linearly on the fibers.

Note that the projection $\pi : (\mathscr{S}, \mathcal{T}(\mathscr{S})) \to \mathscr{S}$ forgets the element in $\mathcal{T}(\mathscr{S})$ and remembers only the distribution itself.

Since each fiber encodes directions of infinitesimal change, while the structure group prescribes how these variations are consistently transported across the manifold, the tangent bundle $\mathcal{T}(\mathscr{S})$ offers a natural language for describing the infinitesimal variations of probability distributions. In this way, statistical learning can be viewed as the geometry of flows on $\mathscr{S}$.

More precisely, tangent spaces are views as centered scores: for $\rho \in \mathscr{S}$ parametrized by θ, the tangent space is spanned by,

$$\mathcal{T}_p(\mathcal{S}) = \text{span}\Big\{\partial_{\theta_i} \ln p(\omega;\theta)\Big\}, \qquad \mathbb{E}_p\big[\partial_{\theta_i} \ln p\big] = 0.$$

Thus, tangent vectors may be read as random variables whose expectation vanishes, encoding the directions in which the distribution "wants" to deform under perturbations.

We consider the Lie group action and transport. The dynamics of learning is mediated by the action of G, where the **linear action** acts on each fiber F_p via G by $v \mapsto A \cdot v$ for $A \in G$. Moreover,

- the **parallel transport** is given along a statistical trajectory $\gamma(t)$; vectors $v \in \mathcal{T}_{\gamma(0)}(\mathcal{S})$ are evolved by the covariant condition

$$\nabla_{\dot\gamma(0)} v = 0 \quad \Longleftrightarrow \quad \dot v^k + \Gamma^k_{ij}\dot\gamma^i v^j = 0,$$

  where Γ^k_{ij} are the Christoffel symbols of the Fisher metric;
- In the **flat case**, for exponential families, G includes translations compatible with ∇.

> In this formulation, the learning process is seen as a trajectory in $\mathcal{S}$, with tangent vectors describing instantaneous directions of adaptation, and parallel transport encoding the memory of how information propagates across the manifold.

In the Lemma below, we are working in the context of Theorem 12.2.

Lemma 13.1 *Consider the fiber bundle* $(\mathcal{T}(\mathcal{S}), \mathcal{S}, \pi, G, F)$.

Let the path $\gamma : I \to \mathcal{S}$ *be a smooth geodesic path in* $\mathcal{S}$*, where* $I \subset \mathbb{R}$*. The fiber over* γ *is denoted by* $F_\gamma = \pi^{-1}(\gamma)$*, which represents the space of tangent vectors along* γ*.*

Then, the fiber

$$F_\gamma = \gamma^+ \sqcup \gamma^{-1}$$

consists of two disjoint connected components. Each component γ^+ *and* γ^- *is contained within a totally geodesic submanifold of* $\mathcal{T}(\mathcal{S})$*, denoted* E^+ *and* E^-*, respectively.*

Proof Consider the fiber above γ. Since for any point of $\mathcal{S}$, its the tangent space is identified to module over paracomplex numbers. This space is decomposed into a pair of subspaces (i.e. eigenspaces with eigenvalues $\pm\varepsilon$). The geodesic curve in $\mathcal{S}$ is a path such that $\gamma = (\gamma^i(t)) : t \in [0, 1] \to \mathcal{S}$, where we have chosen for simplicity $I = [0, 1]$. In local coordinates, the fiber budle is given by $\{\gamma^{ia} e_a\}$, and $a \in \{1, 2\}$. Therefore, the fiber over γ has two components (γ^+, γ^-). Taking the canonical basis for $\{e_1, e_2\}$, implies that (γ^+, γ^-) lie respectively in the subspaces E^+ and E^-. These submanifolds are totally geodesic in virtue of Lemma 3 in [CCN21]. □

We now choose to define a learning process in terms of the Ackley–Hilton–Sejnowski method [AHS85], which is based on minimizing a loss function L.

A loss function L is a smooth function that assigns a real number (a "cost") to every probability distribution ρ_θ in the statistical manifold $\mathscr{S}$. Its purpose is to quantify how "bad" or "incorrect" a model ρ_θ is, with the goal of finding the model that minimizes this cost. The most common and fundamental type of loss function in this geometric setting is a *divergence function*. Specifically, if we have a "true" or "target" data-generating distribution q, the loss is the divergence from q to the model ρ_θ:

$$L(\theta) = D(q \parallel \rho_\theta).$$

Remark 13.2 Warning: $D(q \parallel \rho_\theta)$ is not a distance between two distributions. It is not symmetric: $D(q \parallel \rho_\theta) \neq D(\rho_\theta \parallel q)$.

The most canonical example in our context is the Kullback–Leibler (KL) Divergence. For a target distribution q, the loss is:

$$L_{\mathrm{KL}}(\theta) = D_{\mathrm{KL}}(q \parallel \rho_\theta) = \int q(\omega) \log \frac{q(\omega)}{\rho_\theta(\omega)}\, d\omega.$$

Minimizing this loss is equivalent to performing maximum likelihood estimation.

This process can be interpreted in a geometric framework as follows:

Proposition 13.3 (Geometric Learning Process) *Let $\mathscr{S}$ be a statistical manifold with Fisher metric g. A learning process is a curve $\gamma : I \to \mathscr{S}$ satisfying the* natural gradient flow*:*

$$\dot{\gamma}(t) = -\nabla_g L_{KL}(\gamma(t)),$$

where $L_{KL} : \mathscr{S} \to \mathbb{R}$ is a loss function given by the KL divergence function. The tangent bundle decomposes orthogonally as:

$$T_{\gamma(t)}(\mathscr{S}) = E_t^+ \oplus E_t^-, \quad E_t^\pm := \ker(\mathcal{J}_t \mp \lambda_t Id)$$

where $\mathcal{J}_t$ is the Fisher information matrix at $\gamma(t)$, and λ_t its eigenvalues.

Learning converges when:

$$\lim_{t\to\infty} \|\pi_{E_t^+}\dot{\gamma}(t)\|_g = 0,$$

i.e., when the gradient flow becomes orthogonal to unstable directions.

More precisely, let $\mathscr{S}$ be a *statistical manifold* equipped with a fiber bundle structure $(T(\mathscr{S}), \mathscr{S}, \pi, G, F)$. Consider a geodesic path $\gamma : I \to \mathscr{S}$, parametrized by an interval

$I \subset \mathbb{R}$. The fiber over γ, denoted by F_γ, is assumed to decompose into two connected components:

$$F_\gamma = \gamma^+ \sqcup \gamma^-.$$

Each component γ^+ and γ^- is contained in a *totally geodesic submanifold* of $\mathcal{T}(\mathscr{S})$, denoted by E^+ and E^-, respectively:

$$\gamma^+ \subset E^+, \quad \gamma^- \subset E^-.$$

In particular, the learning process is considered successful whenever the distance between the geodesic γ^+ and its orthogonal projection into E^+ decreases towards zero. That is, the process converges if:

$$d(\gamma^+, \pi_{E^+}(\gamma^-)) \to 0, \quad \text{as the learning iterations progress.}$$

This formulation provides a rigorous geometric criterion for assessing the success of learning, leveraging the underlying differential geometry of the statistical manifold.

In other words:

Proposition 13.4 *The learning process consists in determining if there exist intersections of the paraholomorphic curve γ^+ with the orthogonal projection of the curve γ^- in the subspace E^+.*

The space of centered scores at $P_\theta \in \mathscr{S}$ is:

$$\Upsilon_{P_\theta} = \{X \in L^2(P_\theta) : \mathbb{E}_{P_\theta}[X] = 0, X = g^{-1}(\nabla_\theta \ell \cdot Y) \text{ for } Y \in \mathbb{R}^n\}$$

where $\ell = \ln p(\cdot; \theta)$. For exponential families, $\nabla_\theta \ell \in \Upsilon_{P_\theta}$, and:

$$K(P_{\theta^*}, P_\theta) = \mathbb{E}_{P_\theta^*}[\ell_\theta^*] - \mathbb{E}_{P_\theta^*}[\ell_\theta] = \frac{1}{2}\|\theta - \theta^*\|^2_{g(\theta^*)} + \mathcal{O}(\|\Delta\theta\|^3)$$

The natural gradient flow:

$$\dot{\theta} = -g^{-1}(\theta)\nabla_\theta K(P_{\theta^*}, P_\theta)$$

decomposes as $\dot{\theta} = \dot{\theta}^+ + \dot{\theta}^-$, with convergence when $\|\dot{\theta}^+\|_g \to 0$.

More formally, as was depicted in [BCN99] (sec. 3) let us denote by Υ the set of (centered) random variables over $(\Omega, \mathbf{S}, P_\theta)$ which admit an expansion in terms of the scores under the following form:

$$\Upsilon_P = \{X \in \mathbb{R}^\Omega \mid X - \mathbb{E}_P[X] = g^{-1}(\mathbb{E}_P[Xd\ell]), d\ell\}.$$

By direct calculation, one finds that the log-likelihood $\ell = \ln \rho$ of the usual (parametric) families of probability distributions belongs to Υ_p as well as the difference $\ell - \ell^*$ of log-likelihood of two probabilities of the same family.

Theorem 13.1 (GWS Control of Learning Dynamics) *Let $\mathscr{S}$ be a Frobenius statistical manifold with Fisher metric g, log-partition potential $\tilde{\mathbf{\Phi}}$, and GWS invariants $\tilde{Y}_n$.*

The learning evolution is governed by the gradient flow:

$$\dot{\beta}(t) = -g^{-1}(\beta(t))\nabla_{\beta}L_{KL},$$

where L_{KL} is the Kullback–Leibler (KL) divergence. The GWS invariants constrain this dynamics through:

1. ***Local Convergence**: $\|\nabla_{\beta}L_{KL}\|_g \to 0$ if and only if $\tilde{Y}_2(v,\cdot) = 0$ for all $v \in T_{\beta^*}(\mathscr{S})$;*
2. ***Stability**: The Hessian $\nabla^2 L_{KL}$ is determined by $\tilde{Y}_3$ at critical points;*
3. ***Trajectory Deformation**: $\tilde{Y}_n$ quantify obstruction to flat geodesics.*

Proof **Step 1: Relating L_{KL} to GWS Invariants**. For exponential families, the KL divergence is:

$$\tilde{\mathbf{\Phi}}(\beta) - \langle \beta, \eta^* \rangle - \tilde{\mathbf{\Phi}}(\beta^*)$$

where $\eta^* = \nabla\tilde{\mathbf{\Phi}}(\beta^*)$. Its covariant derivatives at $\beta(0)$ are:

$$\begin{aligned}\nabla_v L_{KL} &= \tilde{Y}_1(v) - \langle v, \eta^* \rangle\\ \nabla^2_{v,w} L_{KL} &= \tilde{Y}_2(v, w)\\ \nabla^3_{u,v,w} L_{KL} &= \tilde{Y}_3(u, v, w)\end{aligned}$$

Step 2: Learning as Gradient Flow. The dynamics $\dot{\beta} = -g^{-1}\nabla L_{KL}$ has convergence determined by:

$$\frac{d}{dt}L_{KL} = -\langle \nabla L_{KL}, \nabla L_{KL}\rangle_g = -\|\nabla L_{KL}\|_g^2$$

Vanishing occurs if and only if $\nabla L_{KL}|_{\beta^*} = 0$, equivalent to $\tilde{Y}_1(v) = \langle v, \eta^* \rangle$ for all v.

Step 3: GWS Control of Stability. At critical points, the Hessian satisfies:

$$g(\nabla_v^2 K(w), u) = \tilde{Y}_2(v, w, u)$$

Stability requires $\tilde{Y}_2(v,\cdot)$ positive definite on $T_{\beta^*}(\mathscr{S})$.

Step 4: WDVV Constraints. When $\mathscr{S}$ is Frobenius, the associativity equations:

$$\sum_{\alpha,\beta} g^{\alpha\beta}\tilde{Y}_3(e_i, e_j, e_\alpha)\tilde{Y}_3(e_k, e_l, e_\beta) = \sum_{\alpha,\beta} g^{\alpha\beta}\tilde{Y}_3(e_i, e_k, e_\alpha)\tilde{Y}_3(e_j, e_l, e_\beta)$$

enforce compatibility between learning directions, preventing degenerate convergence. □

Similarly as in the classical (GW) case, the (GWS) count intersection numbers of the para-holomorphic curves generated by γ^+ and γ^-. In fact, we have the following statement:

Corollary 13.1 (GWS as Enumerative Learning Invariants) *Let $\gamma : [0, 1] \to \mathscr{S}$ be a learning trajectory. The GWS invariants count:*

- ***Constraint Satisfaction**: Virtual intersections with statistical hypersurfaces $\mathcal{Z}_j(c_j)$*
- ***Transition States**: Critical points along the learning path*
- ***Stability Transitions**: Bifurcations in the gradient flow*

Specifically, for the pullback bundle $\gamma^*\mathcal{T}(\mathscr{S})$:

$$\langle \mathcal{Z}_{j_1}, \ldots, \mathcal{Z}_{j_n} \rangle_{0,n}^{\gamma} = \int_{\overline{\mathcal{M}}_{0,n}(\gamma)} \prod_{k=1}^{n} \mathrm{ev}_k^*[\mathcal{Z}_{j_k}] = \tilde{Y}_n(v_1, \ldots, v_n)$$

where $\overline{\mathcal{M}}_{0,n}(\gamma)$ is the moduli space of learning curves deforming γ with n constraint evaluations.

Remark: The proof is omitted from this book. This proof uses techniques from equivariant geometry and Gromov–Witten theory. The key idea is that the moduli space of learning curves deforming γ can be analyzed locally via the normal bundle, and the intersection numbers are encoded in the $\mathbb{C}^*$-weights.

13.6 Final Word

This work shows that the geometry of statistical models can be rigorously connected to the topology of learning dynamics and statistical Gromov–Witten invariants. Via our new interpretation of learning trajectories, we gain a precise mathematical construction to count and classify critical points, transitions, and bifurcations in gradient flows in statistical models. This framework opens a novel pathway to understand inference through geometric and topological methods, revealing deep structural insights into the landscape of learning.

References

AHS85. Ackley, D., Hilton, G., & Sejnowski, T. (1985). Learning algorithm for Boltzmann machine. *Cognitive Science, 9*, 147–169.

Am85. Amari, S. *Differential-geometrical methods in statistics.* Lecture Notes in Statistics, **28**. Berlin: Springer-Verlag (1985).

Am97. Amari, S. (1997). Information Geometry. In J.-P. Bourguignon & H. Nencka (Eds.), *Contemporary Mathematics* (Vol. 203, pp. 81–95). Providence, RI: American Mathematical Society. Geometry and Nature.

Aus64. Auslander, L. (1964). The structure of complete locally affine manifolds. *Topology, 3*(Supplement 1), 131–139.

Bnz60. Benzécri, J.-P. (1960). *Sur les variétés localement affines et localement projectives* (pp. 229–332) Bulletin de la S. M. F., tome 88.

Bb11. Bieberbach, L. (1911). Über die Bewegungsgruppen der Euklidischen Räume. *Erste Abhandlung.) Mathematische Annalen, 70*, 297–336.

Bb12. Bieberbach, L. (1912). Über die Bewegungsgruppen der Euklidischen Räume. (Zweite Abhandlung.) Die Gruppen mit einem endlichen Fundamentalbereich. *Mathematische Annalen, 72*, 400–412.

Bl95. Billingsley, P. (1995). Probability and Measure. 3rd Edn. Wiley Series in Probability and Mathematical Statistics. Wiley.

BCN99. Burdet, G., Combe, Ph., Nencka, H. (1999). Statistical manifolds, self-parallel curves and learning processes. In *Seminar on Stochastic Analysis, Random Fields and Applications (Ascona, 1996)* (pp. 87–99). Progress in Probability, **45**. Birkhäuser

BCN00. Burdet, G., Combe, Ph., Nencka, H. (2000). Statistical manifolds: the natural affine-metric structure of probability theory. In *Mathematical Physics and Stochastic Analysis (Lisbon, 1998)*, (pp. 138–164). World Scientific.

BN97. Burdet, G., & Nencka, H. (1997). Equation of self-parallel curve deviation on statistical manifolds. *Methods of Functional Analysis and Topology, 3*(1), 46–50.

BNP97. Burdet, G., Nencka, H., & Perrin, M. (1997). An example of dynamical behaviour of the relative entropy. In J.-P. Bourguignon, & H. Nencka (Eds.), *Geometry and nature contemporary mathematics* (Vol. 203, pp. 97–104). American Mathematical Society

© The Editor(s) (if applicable) and The Author(s), under exclusive license to Springer Nature Switzerland AG 2026

N. C. Combe et al., *Exploring Information Geometry*, Frontiers in Mathematics,
https://doi.org/10.1007/978-3-032-23054-6

Ca54. Calabi, E. (1954). On the space of Kähler metrics. *Proceedings of the National Academy of Sciences, 40*(8), 759–760.

CDM93. Carrière, Y., Dal'bo, F., & Meigniez, G. (1993). Inexistence de structures affines sur les fibrés de Seifert. *Mathematische Annalen, 296*, 743–753.

Che82. Chentsov, N. (1982). Statistical decision rules and optimal inference. Translations of mathematical monographs, 53, American Mathematical Society.

Chq54. Choquet, G. (1954). Theory of capacities. *Annales de l'institut Fourier, tome, 5*, 131–295.

C18. Combe, N. (2018). Geometric Classification of Real Ternary Octahedral Quartics. *Discrete Computational Geometry, 60*(2), 255–282.

CCN22. Combe, N.C., Combe, P., & Nencka, H. (2022) Algebraic properties of the information geometry's fourth Frobenius manifold. In *Advances in information and communication, lecture notes in networks and systems* (Vol. 438), Springer

CM20. Combe, N. C., & Manin, Y. I. (2020). F-manifolds and geometry information. *Bulletin of the London Mathematical Society, 52*(5), 777–792.

CCN21. Combe, N. C., Combe, P., & Nencka, H. (2021). Frobenius statistical manifolds and geometric invariants. Lecture Notes in Computer ScienceIn F. Nielsen & F. Barbaresco (Eds.), *Geometric science of information–5th international conference, GSI 2021, Paris, France, July 21–23, 2021, Proceedings* (Vol. 12829, pp. 565–573). Springer.

CCN22. Combe, N. C., Combe, P., & Nencka, H. (Eds.) (2022). Statistical frobenius manifolds, Lie groups and learning. XXVIIIe Colloque GRETSI—Traitement du Signal et des Images, Springer

CN97. Combe, P., & Nencka, H. (1997). Information geometry and learning in formal neural networks. In Bourguignon, J.-P., Nencka, H. (Eds.), *Geometry and Nature. Contemporary Mathematics* (Vol. 203, pp. 105–116). American Mathematical Society

DM75. Dellacherie, C., & Meyer, P.-A. (1975). *Probabilités et potentiel, Chapitres 1–4*. Hermann.

D96. Dubrovin, B. (1993). Integrable systems and quantum groups. *Lecture Notes in Mathematics, 1620*, 120–348.

FK94. Faraut, J., & Koranyi, A. (1994). *Analysis on symmetric cones*. Oxford Mathematical Monographs.

Fe66. Feller, W. (1966). *An introduction to probability theory and its applications*. John Wiley & Sons.

FG83. Fried, D., & Goldman, W. M. (1983). Three-dimensional affine crystallographic groups. *Advances in Mathematics, 47*(1), 1–49.

G22. Goldman, W.M. (2022). *Geometric Structures on Manifolds*. Graduate Studies in Mathematics, vol. 227, American Mathematical Society, Providence, RI.

G85. Gromov, M. (1985). Pseudo holomorphic curves in symplectic manifolds. *Inventiones Mathematicae, 82*, 307–347.

JM81. Jimbo, M., & Miwa, T. (1981). Monodromy perserving deformation of linear ordinary differential equations with rational coefficients. *II. Physica D: Nonlinear Phenomena, 2*(3), 407–448.

KaW96. Kadane, J. B., & Wasserman, L. (1996). Symmetric, coherent, Choquet capacities. *The Annals of Statistics, 24*(3), 1250–1264.

KaS90. Kashiwara, M., & Schapira, P. (1990). Sheaves on Manifolds. Grundlehren der mathematischen Wissenschaften (Vol. 292). Springer-Verlag.

Ko79. Kobayashi S. (1979). On holomorphic connections. In A. Dold & B. Eckmann (Eds.), Global differential geometry and global analysis, Vol. 838, *Proceedings of the Colloquium Held at the Technical University of Berlin*, November 21–24

KobNo96. Kobayashi, S., & Nomizu, K. (1996). *Foundations of differential geometry. Wiley Classics Library Edition*. John Wiley & Sons.

Koe69. Koecher, M. (1969). *An elementary approach to bounded symmetric domains. Lecture Notes.* Rice University.

KonMa94. Kontsevich, M., & Manin, Y. I. (1994). Gromov-witten classes, quantum cohomology, and enumerative geometry. *Communications in Mathematical Physics, 164*(3), 525–562.

Ku72. Kuratowski, K. (1972). *Introduction to set theory and topology.* Pergamon Press: PWN - Polish Scientific Publishers.

La95. Lang, S. (1995). *Differential and Riemannian manifolds.* Springer-Verlag.

La99. Lang, S. (1999). *Fundamentals of differential geometry. Graduate Texts in Mathematics.* Springer-Verlag.

Le94. Letac, G. (1994). Les familles exponentielles statistiques invariantes par les groupes du Cône et du paraboloïde de révolution. Journal of Applied Probability, 31, Studies in Applied Probability.

LiZu93. Lian, B. H., & Zuckerman, G. (1993). New perspectives on the BRST-algebraic structure of string theory. *Communications in Mathematical Physics, 154*, 613–646.

LV03. Lott, J., & Villani, C. (2009). Ricci Curvature form a metric-measure spaces via optimal transport. *Annals of Mathematics, 169*, 903–991.

MC04. McCrimmon, K. (2004). *A Taste of Jordan Algebras.* Springer: Universitext.

Ma99. Manin, Yu. I. *Frobenius manifolds, quantum cohomology, and moduli spaces.* Colloquium Publications, **47**. Providence, RI: American Mathematical Society (1999).

Me68. Meyer, P.-A. (1968). *Guide détaillé de la théorie*<< générale >>*des processus.* Séminaire de probabilités (Strasbourg), tome 2, pp. 140–165.

Mi65. Milnor, J. W. (1965). *Topology from the differentiable viewpoint.* University Press of Virginia.

MoCh89. Morozova, E. A., & Chentsov, N. N. (1991). Markov invariant geometry on state manifolds. *Journal of Soviet Mathematics,56*(5), 2648–2669. Russian original 1989.

MoCh91-1. Morozova, E. A.; Chentsov, N. N. *Projective Euclidean geometry and noncommutative probability theory.Trudy Matematicheskogo Instituta imeni V. A. Steklova*, **196**, 105–113 (1991); Proceedings of the Steklov Institute of Mathematics, **196**, 117–127 (1992).

MoCh91-2. Morozova, E. A., & Chentsov, N. N. (1991). Natural geometry of families of probability laws. Itogi Nauki i Tekhniki. *Seriya Sovremennye Problemy Matematiki. Fundamental'nye Napravleniya, 83*, 133–265.

No47. Norden, A. P. (1947). La connexion affine sur les surfaces de l'espace projectif. *Rec. Math. [Mat. Sbornik] N.S, 20*(62), 263–281.

No52. Norden, A. P. (1952). Biaffine space and its mapping on itself, (Russian) Kazan. *Gos. Univ. Uč. Zap., 112*(10), 3–11.

No54. Norden, A. P. (1954). On a complex representation of tensors of a biplanar space, (Russian) Kazan. *Gos. Univ. Uč. Zap., 114*(8), 45–53.

No58. Norden, A. P. (1958). On intrinsic geometry of the second kind on hypersurfaces of an affine space. *Izvestiya Vysshikh Uchebnykh Zavedenii. Matematika, 4*(5), 172–183.

No63. Norden, A. P. (1963). Cartesian composition spaces. *Izvestiya Vysshikh Uchebnykh Zavedenii. Matematika, 35*(4), 117–128.

Pa67. Parthasarathy, K. R. (1967). *Probability measures on metric spaces.* Academic Press.

Pe81. Peirce, B. (1881). Linear associative algebra. *American Journal of Mathematics, 4*, 97–229.

Ro97. Rozenfeld, B. (1997). *Geometry of Lie groups.* Springer-Science+Business Media.

RuTi95. Ruan, Y., & Tian, G. (1995). A mathematical theory of quantum cohomology. *Journal Differential Geometry, 42*(2), 259–367.

Sa07. Sabbah, C. (2008). *Isomonodromic deformations and Frobenius manifolds: An Introduction Textbook.* London.

Sh54. Shirokov, A. P. (1954). The geometry of generalized biaxial spaces, (Russian) Kazan. *Gos. Univ. Uč. Zap., 114*(2), 123–166.

Sh87. Shirokov, A. P. (1987). Affine connection spaces (some aspects of the method of normalization of A.P. Norden's). *Journal of Soviet Mathematics, 37*, 1239–1253.

Sh02. Shirokov, A. P. (2002). Spaces over algebras and their applications, Geometry, 7. *Journal of Mathematical Sciences (New York), 108*(2), 232–248.

Shu77. Shurygin, V. V. (1977). Totally geodesic subspaces in spaces over algebras, (english translation) Izvestiya VUZ. *Matematika, 21*(11), 127–129.

Sik71. Sikorski, R. (1971). Differential modules. *Colloquium Mathematicum, 24*(1), 45–79.

Sm81. Smillie, J. (1981). An Obstruction to the Existence of Affine Structures. *Inventiones mathematicae, 64*, 411–416.

Szcz24. Szczepański, A. *Geometry of Crystallographic Groups*, Algebra and Discrete Mathematics: Volume 5, 2nd Edition (2024)

Vil03. Villani, C. (2003). *Optimal transport, graduate studies in mathematics*. (Vol. 58). Providence, Rhode Island: American Mathematical Society.

V60. Vinberg, E. B. (1960). Homogeneous cones. *Doklady Acad. Nauk. USSR, 133*, 9–12.

V63. Vinberg, E. B. (1963). The theory of homogeneous convex cones. *Transactions of the Moscow Mathematical Society, 12*, 340–403.

V65. Vinberg, E. B. (1965). The structure of the group of automorphisms of a homogeneous convex cone. *Transactions of the Moscow Mathematical Society, 14*, 63–93.

W91. Witten, E. (1991). Two-dimensional gravity and intersection theory on moduli space. *Surveys in Differential Geometry, 1*, 243–310.

Index

A
Absolute continuity, 111, 112, 152, 153
Affine connection, 95, 97, 114, 145–147, 164, 167, 175, 178, 179, 185, 205, 229
Affine coordinates, 77, 160, 164–166, 183, 190, 216
Affine geodesic, 81
Affine structure, 173, 175, 178, 179, 183, 186, 190
α-connection, 116, 120, 122
Amari–Chentsov tensor, 116, 117
Application
  continuity, 14
  homéomorphism, 14
Atlas, 30, 34, 44, 177, 184

B
Baire space, 15, 16, 26
Banach manifold, 34, 39, 45, 46, 154, 155, 232, 235
Banach space, 33, 34, 37, 39
Bianchi identity, 98–100
Borel σ-algebra, 108, 110
Bundle, 59
  cotangent bundle, 67
  cross-section, 62
  fiber bundle, 59
  G-bundle, 61
  moving frame, 67
  principal fiber bundle, 62
  tangent bundle, 63
  trivial bundle, 61
  vector bundle, 61, 62

C
Canonical parameter, 248
Caph (space of probability measures), 152
Category, 6, 16, 26, 85–88, 128, 131, 134, 139, 140, 142, 143, 147, 151, 156
Ceva line, 162, 164, 165
Chart, 30–32, 34, 39, 44, 45, 47, 48, 50, 51, 53, 65, 80, 83, 156–159, 167, 175, 177
Chentsov uniqueness theorem, 133
Christoffel symbols, 75, 76, 96, 116, 122, 180, 203, 257
Compactness, 4, 9, 16
Connectedness, 9, 17
Connection
  christoffel symbols, 75
  coefficients, 73
  connection forms, 73
  linear, 72
  parallel transport, 80
Connection coefficients, 73, 75, 76, 96, 98, 116, 181, 202, 206
Connection forms, 73
Contractible space, 22
Cotangent bundle, 67, 68, 70, 237, 238
Cotangent space, 51
  finite dimensional, 52
Covariant derivative, 5, 71–73, 75, 77–80, 87, 88, 96, 99, 116, 147, 185, 190, 203, 248–250, 260

© The Editor(s) (if applicable) and The Author(s), under exclusive license to Springer Nature Switzerland AG 2026

N. C. Combe et al., *Exploring Information Geometry*, Frontiers in Mathematics,
https://doi.org/10.1007/978-3-032-23054-6

Covector, 52
Covector field, 68
Covering space, 18, 19, 27, 196
Cramér-Rao bound, 123
Cross-section of a bundle, 62
Crystallographic group, 176, 177
Curvature, 3, 29, 41, 59, 71, 72, 76, 83, 84, 87–90, 93–98, 100, 120, 146, 149, 151, 154, 183, 201–207, 238, 239, 243

D
Darboux–Egoroff coordinates, 180, 181, 196
Deep learning, 6
Derivation, 42, 45, 49, 50, 66, 160
Derivative
  derivation, 49
  directional, 49
Diffeomorphism, 45, 69
  image of a covector field, 69
  image of a vector field, 69
Differentiable manifold, 1, 29, 41, 42, 44, 46–49, 64, 67, 69, 72, 73, 205, 218
Differentiable mapping, 45, 69, 70
Differential, 54
Directional derivative, 43, 44, 48, 49, 83, 117
Disjoint sum topology, 23
Divergence function, 258
Dominated family, 143
Dominating measure, 111, 113, 131, 144, 160
Dual connection, 242

E
$\mathcal{E}$-modeled manifold, 32
Entropy, 246
Equivariant connection, 147, 148
Erlangen program, 125, 131, 149
Euler vector field, 189, 196, 199, 200
Expectation, 111, 159, 237, 238, 257
Exponential family, 113–115, 123, 124, 159, 235, 237, 239, 247, 248

F
Fiber bundle, 5, 59–63, 70, 84, 257, 258
Fibration, 204
Fisher information, 114, 121, 132, 133, 237, 248, 258
Fisher-Rao metric, 114, 115, 117, 235
Flat connection, 146, 167, 186, 228, 241, 243
Flat coordinates, 180–184, 190, 239, 240, 247
Foliation, 100, 204–206, 238
Frame, 51
Frame bundle, 62–64
Fréchet space, 38
Frobenius algebra, 188, 189, 191–195, 198
Frobenius manifold, 1, 4, 6, 77, 173–175, 180, 181, 187–189, 191, 194, 196, 198–201, 204–207, 239–241, 246, 247, 250
Functor, 85, 86, 88, 129–131, 139, 140, 144, 148
Fundamental group, 19, 20, 27, 176, 179, 213

G
Gâteaux derivative, 5, 41–43
Gaussian distribution, 114, 131, 132, 144
G-bundle, 62
Geodesic, 80–83, 96–98, 114, 122, 145–149, 151, 154, 159, 241–243, 257–260
  affine, 81
Geodesic midpoint, 148
Gromov–Witten invariants, 250, 253, 256

H
Hausdorff space, 13–16, 25
Hilbert manifold, 34, 46
Hilbert space, 33, 34, 39, 46
Homeomorphism, 9, 14, 16, 18, 19, 23, 24, 30–32, 34, 61, 156
Homogeneous space, 64, 145
Homological mirror symmetry, 174
Homotopy, 21, 22, 146
Hypersurface, 157, 183, 217, 246, 248–252, 254, 255, 261

I
Idempotent, 195, 196, 198–201, 206, 220–223, 226–228, 242, 243
Implicit function theorem, 37, 45

Information geometry, 1, 4–6, 85, 95, 103, 113, 114, 119, 122, 127, 131, 151, 220, 225, 245, 246
Isomonodromy, 198–200

J
Jordan algebra, 225–229
Jordan decomposition, 112

K
Klein figure, 127, 128, 131, 132
Klein geometry, 145
Kullback–Leibler divergence, 114, 258, 260

L
Lebesgue measure, 25, 109, 114, 133, 153, 230
Levi–Civita connection, 83, 96, 97, 119, 122, 147, 180, 183, 185, 197, 198, 202, 203, 206, 239, 240, 242
Lie algebra, 66
  Jacobi's identity, 67
  lie bracket, 66
Lie bracket, 65–67, 83, 87, 88, 96, 99, 100, 146, 203, 205
Lie group, 29, 42, 64, 256, 257
Linear connection, 72
Locally convex space, 34, 45
Log-partition function, 113, 116, 160, 248–251

M
Machine learning, 1, 3, 41, 95
Manifold
  atlas, 30
    admisible, 44
    compatible, 44
  Banach, 45
  chart, 30
  curve, 48
  differentiable, 41, 44
  finite manifold
    differentiable structure, 47
    local coordinate system, 48
  modeled manifold, 32
  Riemann, 46
  tensor at a point, 55
  topological manifold, 30
Mapping
  diffeomorphism, 45
  differentiable, 45
Markov kernel, 6, 125, 127, 133, 136–144, 151, 154, 155
Markov morphism, 131–133, 138–140, 142–144
Markov transformation, 132, 139
Maximum likelihood estimation, 122, 123, 258
Measurable function, 110, 112, 139, 141, 155, 159, 170, 231
Measure, 3, 5, 6, 25, 88, 94, 96, 98, 100, 103–105, 107–109, 111–116, 119, 121, 125, 126, 131–144, 151–161, 163, 165, 168–170, 189, 206, 215, 230–232, 234, 235, 239
Metric space, 9, 11, 15, 16, 25, 26, 46
Mirror symmetry, 174
Modeled manifold, 29, 32–34
Moduli space, 173, 174, 247, 249, 251–253, 255, 261
Monge–Ampère equation, 183, 185–187
Monodromy, 199–201
Moving frame, 67, 79
m-pair, 211, 217–219
Multinomial distribution, 107, 132
Musical isomorphism, 117

N
Natural gradient, 258, 259
Natural parameter, 113, 124, 237, 238, 254
Natural transformation, 87, 148
Neural network
Normed vector space, 16, 38, 45

P
Paracomplex algebra, 221, 228, 241
Paracomplex Dolbeault lemma, 236
Paracomplex geometry, 230, 232, 233
Para-Kähler geometry, 237
Parallel transport, 3, 5, 71, 72, 76, 80, 83, 95, 116, 146, 147, 185, 190, 228, 229, 257
Parametric family, 113, 134
Partition function, 165, 166, 247, 248

Path-connected space, 20
Peirce decomposition, 226–228, 243
Peirce reflection, 228, 241–243
Poisson distribution, 108
Potential function, 188, 189, 246, 247, 250
Pre-Frobenius manifold, 174, 183–187, 190, 191, 194, 201, 204, 209
Pre-Lie algebra, 179, 180, 209
Prepotential, 181
Presheaf, 84–86
Principal fiber bundle, 62, 63
Probability distribution, 3, 4, 6, 104, 106, 107, 114, 115, 120, 121, 125–127, 131–135, 151, 160, 174, 191, 225, 228, 230, 236, 249, 250, 256, 258, 259
Probability measure, 3, 6, 104, 105, 107–109, 111–113, 131–134, 136–138, 141, 143, 144, 151–155, 157, 160, 168, 170, 230–232, 239
Probability space, 3, 104–106, 110, 136
Projective geometry, 211, 212, 214, 218, 219, 222, 230
Projective space, 20, 30, 158, 211, 212, 214, 217–219, 222, 229, 241

R

Radon–Nikodym theorem, 112, 113
Random variable, 110, 111, 115, 121, 151, 257, 259
Rectilinear coordinate system, 159
Relative entropy
Riemannian manifold, 29, 46, 47, 83, 87, 147, 176, 177
Riemannian metric, 29, 64, 115, 132, 133, 166, 167, 234, 235

S

Sample space, 3, 103, 104, 106–108, 133, 135, 158, 161
Score function, 256
Second fundamental form, 147, 242
Sectional curvature, 93–95
Semisimple Frobenius manifold, 194, 196, 198, 204, 206, 207
Set
  covering, 16
    compact, 16
    open, 16
    relatively compact, 16
    subcovering, 16
  separable, 15
  subset
    adherence, 15
    closed, 15
    closure, 15
    dense, 15
    locally compact, 16
    nowhere dense, 15
Sheaf, 5, 42, 71, 84, 86–89, 91, 92, 97, 179, 180, 184, 196
σ-algebra, 3, 4, 103–105, 107–110, 131, 134–137, 139, 140, 156, 157
Signed measure, 111–113, 153, 161, 163, 230

Simply connected space, 19
Space
  connectivity, 17
  Hausdorff, 13
  locally connected, 17
  normal, 13
  paracompact, 16
Spencer differential, 100
Statistical manifold, 113, 116, 117, 119, 121, 122, 131–133, 151, 154, 156, 160, 161, 164, 167, 191, 194, 211, 220, 225, 228, 245, 247–254, 256, 258–260
Structure connection, 190, 191, 201, 204
Submanifold, 32, 39, 40, 93, 114, 133, 146, 147, 158, 205, 208, 234, 239, 241–243, 248, 257, 259
Sufficient statistic, 113, 124, 247–251, 254

T

Tangent bundle, 32, 63–65, 67, 70, 84, 87, 88, 90, 91, 97, 100, 116, 175, 179, 188, 197, 232, 238, 256, 258
Tangent space, 49
Tangent vector, 48
  differentiable manifold, 49
  tangent vector to a curve, 48
Tensor, 54
  symmetry properties, 58
  tensor product, 56
Tensor field, 70, 72, 78, 166
Tensor product, 54–56, 80, 88–92, 175

Topological manifold, 5, 29, 30, 44
Topological space, 5, 9, 10, 12–20, 22, 23, 25, 29, 32, 34, 47, 59–61, 84–86, 105
Topology
  coarse, 12
  compactification, 16
  discrete, 12
  fine, 12
  limit point, 15
  theorem
    Bolzano-Weierstrass, 16
    Hein-Borel, 16
  trivial, 12
Torsion tensor, 95–98, 100, 146
Totally geodesic submanifold, 146
Translation-invariant tensor field, 166
Trivial bundle, 61

U
Universal covering, 19, 20, 27
Universal property, 24, 54, 55, 88, 90, 92, 134, 138, 140

V
Vector bundle, 39, 59, 62, 64, 68, 83, 84, 87, 99, 100, 116
Vector field, 65
  invariant vector field, 69

W
WDVV equation, 4, 173, 174, 188, 194, 239–241, 250
Web structure, 204–206

MIX
Papier | Fördert gute Waldnutzung
FSC® C083411

Zeitfracht Medien GmbH
Ferdinand-Jühlke-Straße 7
99095 Erfurt, Deutschland
produktsicherheit@kolibri360.de